Honeybee Ecology

MONOGRAPHS IN BEHAVIOR AND ECOLOGY

Edited by John R. Krebs and
Tim Clutton-Brock

Five New World Primates: A Study in Comparative
Ecology, *by John Terborgh*

Reproductive Decisions: An Economic Analysis of
Gelada Baboon Social Strategies, *by R. I. M. Dunbar*

Honeybee Ecology: A Study of Adaptation in Social Life,
by Thomas D. Seeley

Honeybee Ecology

A Study of Adaptation in Social Life

THOMAS D. SEELEY

Princeton University Press
Princeton, New Jersey

Library of Congress Cataloging in Publication Data will
be found on the last printed page of this book

ISBN 0-691-08391-6
ISBN 0-691-08392-4 (pbk.)

This book has been composed in Linotron Times Roman
and Univers 45

Clothbound editions of Princeton University Press books
are printed on acid-free paper, and binding materials are
chosen for strength and durability. Paperbacks, although
satisfactory for personal collections, are not usually
suitable for library rebinding

Printed in the United States of America by
Princeton University Press, Princeton, New Jersey

Dedicated to the honeybees
living in the forests around Ithaca, New York

Contents

Acknowledgments   ix

1   Introduction   3

    Why Study the Ecology of Honeybees?   3
    Individual-Level versus Colony-Level Selection   5

2   Honeybees in Nature   9

    Fossil Honeybees   9
    Systematics of Living Honeybees   10
    "Domestication" of Honeybees   14
    The Importance of Studying Honeybees
      in Nature   17

3   The Honeybee Societies   20

    Kin Structure of Colonies   20
    Queens and Workers: Reproductive Division
      of Labor   22
    Labor Specialization by Workers   31
    Colony Life Cycle   36

4   The Annual Cycle of Colonies   39

    Introduction   39
    Annual Cycle of Energy Intake
      and Expenditure   40
    Annual Cycles of Colony Growth
      and Reproduction   43
    Evolution of the Annual Cycle   46

5   Reproduction   49

    Introduction   49
    Investment Ratio between Queens and Drones   49
    Paternity of Virgin Queens   54
    Life History Traits   57
    Mating System   67

6   Nest Building   71

    Selecting a Nest Site   71
    Comb Construction   76

7   Food Collection   80

    Introduction   80
    Colony Economics   81

Recruitment to Food Sources 83
The Information-Center Strategy of Foraging 88
Decision Making by Colonies 93
Decisions by Individuals in a Flower Patch 103

8 Temperature Control 107

Sociality and the Origins of
Nest Thermoregulation 107
Benefits of Nest Temperature Control 110
Heating the Nest 111
Cooling the Nest 116
Thermoregulation in Swarms 118
Thermoregulation during Foraging 121

9 Colony Defense 123

Evolutionary Perspectives 123
Catalogue of Protective Mechanisms 125
Cost-Benefit Analyses of Defense 134

10 Behavioral Ecology of Tropical
Honeybees 138

The Importance of Studies in the Tropics 138
African and European Versions of *Apis mellifera* 139
Asian *Apis*: Contrasts in Adaptation among
Closely Related Species 149

Literature Cited 161

Author Index 193

Subject Index 197

Acknowledgments

I wish to express here my gratitude to the numerous colleagues whose intellectual stimulation has helped me begin to understand the honeybee societies from an evolutionary perspective. A partial list includes: Bernd Heinrich, Bert Hölldobler, Nancy Knowlton, Roger A. Morse, Richard Nowogrodzki, P. Kirk Visscher, and Edward O. Wilson. I also feel deeply grateful to have had the opportunity to spend the past fourteen years in three quite different, but complementary, intellectual environments: first the Dyce Laboratory for honeybee studies at Cornell, then the social insect group at Harvard, and finally the ecology and evolution division at Yale. It is a pleasure to acknowledge my debt to these groups of individuals.

I also owe deep thanks to Fred C. Dyer, Christine A. Evers, Bernd Heinrich, Charles D. Michener, Roger A. Morse, Francis Ratnieks, Gene E. Robinson, David W. Roubik, Robin Hadlock Seeley, Randy Thornhill, and Mark L. Winston, for critical readings of various parts of the manuscript. Their comments have helped me greatly. Laurie Burnham, Iwao Kudo, Gene E. Robinson, David W. Roubik, Friedrich Ruttner, Shôichi F. Sakagami, and Paul Schmid-Hempel gave me unpublished manuscripts, original photographs, or other forms of aid. To all I am grateful.

Special thanks must be given to the artists who created the illustrations for this book. Sandra Olenik prepared all the original drawings (including the cover drawing) and redrew all the previously published figures, while Helen Flynn did the typesetting for the drawings, and William K. Sacco produced all the photographs.

Permission to reproduce certain illustrations was granted by the following persons: Leslie Bailey, Laurie Burnham, Harald Esch, Wayne M. Getz, James L. Gould, Bernd Heinrich, Henry S. Horn, Martin Lindauer, Hermann Martin, Charles D. Michener, Walter C. Rothenbuhler, Friedrich Ruttner, Paul Schmid-Hempel, Edward E. Southwick, P. Kirk Visscher, Edward O. Wilson, and Mark L. Winston. The following publishers gave permission to reproduce illustrations for which they hold the copyright: Akademische Druck- u. Verlagsanstalt, American Society of Zoologists (*American Zoologist*), Blackwell Scientific Publications (*Ecological Entomology* and Krebs and Davies' *Behavioral Ecology. An Evolutionary Approach*), Company of Biologists Ltd. (*Journal of Experimental Biology*), Cornell University Press, Ecological Society of America (*Ecology* and *Ecological Monographs*), W. H. Freeman and Co. (*Scientific American*), Harvard University Press, International Bee Research Association (*Journal of Apicultural Research* and Dade's *Anatomy and Dissection of the Honeybee*), Macmillan Journals Ltd. (*Nature*), Masson S.

A. (*Insectes Sociaux*), Pergamon Press (*Comparative Biochemistry and Physiology*), Springer Verlag (*Zeitschrift für vergleichende Physiologie* and *Behavioral Ecology and Sociobiology*). A few pages of an article by Seeley and Heinrich (1981) have been repeated with little change in Chapter Eight of this book; I thank John Wiley and Sons, Inc. for permission to use this material.

Finally, I wish to acknowledge my indebtedness to Yale College for a fellowship which gave me the year without distractions which I needed to write this book, and to several organizations—the National Science Foundation, National Geographic Society, American Philosophical Society, and Society of Fellows at Harvard—whose financial support has enabled me to investigate the behavioral ecology of honeybees in several different parts of the world.

Thomas D. Seeley
New Haven, Connecticut
17 November 1984

Honeybee Ecology

1 Introduction

Why Study the Ecology of Honeybees?

The honeybee is a wonderful example of adaptation. In this it resembles all forms of life, but because it is an extremist its adaptations are striking. The honeybee's waggle dance, with which forager bees share information about the locations of new patches of flowers, is unsurpassed among animal communication systems in its capacity for coding precise yet flexible messages. Honeybee workers display an extraordinarily elaborate division of labor by age, switching their labor roles at least four times as they grow older. When a honeybee colony needs a new home, several hundred scout bees comb some 100 square kilometers of forest, discover a few dozen possible nest cavities, and harmoniously choose the best dwelling place through a sort of plebiscite. In winter, the thousands of honeybees in a colony form a tight, well-insulated cluster and pool their metabolic heat—fueled by about 20 kilograms of honey stores—to keep warm despite subfreezing temperatures, a method of winter survival which is unique among insects. The honeybee, then, has an extremely elaborate social life. It is therefore an unusually rewarding subject for ecological studies of social behavior.

Besides possessing a wealth of adaptations associated with group living, the honeybee's attractiveness for ecological investigation is heightened by the remarkable ease with which it is studied. Honeybee colonies thrive as managed colonies in apiaries or as wild colonies in nature, or both, throughout most regions of the world. Unlike most other social insects, honeybees prosper in brightly illuminated, glass-walled nests and so allow humans to observe easily the internal operations of their societies. Furthermore, individual honeybees are relatively large social insects, large enough so that colony members can be labelled with color codes for individual identification. This sets the stage for truly detailed observations of interactions among colony members. Countless experimental manipulations of the honeybee's social environment, such as colony fusions, brood transplantations, and alterations of nest design, were made possible by the invention of hives with movable combs in the mid 1800's. Nests which are readily dissected and reassembled also facilitate the collection of such basic ecological data as colony population size, age structure, and metabolic rate. Even manipulations of the kinship relations among colony members are possible with honeybees, a fortuitous byproduct of instrumental insemination techniques developed for bee breeding.

Given the richness of the honeybee's adaptations for social life and its advantages as a study animal, it is somewhat surprising that a strong imbalance exists between mechanistic and functional studies of honeybee sociality. We know a great deal about how honeybee societies work but comparatively little about the forces of natural selection which have shaped their finely tuned social systems. Perhaps the most vivid illustration of this imbalance is found in our understanding of the honeybee's social organization for food collection. The central mechanism of their foraging strategy is recruitment of nestmates by successful foragers via the dance language. The physiological processes underlying the dance language have been the prime subject of investigation by several dozens of researchers over four decades, and the dance is certainly one of the best understood examples of animal behavior. In contrast, our knowledge of how foraging efficiency is enhanced by this social machinery is still nascent. Undoubtedly there are numerous reasons for this difference in emphasis between physiological and ecological approaches. In part, it is a reflection of the history of scientific studies on animal behavior, which focused first on questions of behavioral mechanisms and only relatively recently on topics of behavioral ecology. Perhaps more importantly, though, it reflects the ease with which one can culture honeybees in apiaries and perform experiments with them. Thus honeybee scientists seem to have been consistently attracted to experimental studies conducted in man-made environments, rather than broadly observing the organism living undisturbed in nature, the essential first stage of behavioral-ecological studies.

This book is an attempt to redress the imbalance between physiological and ecological studies of honeybee social life by emphasizing ecological studies of the honeybee societies. It will focus on how honeybees live in nature and why their social organization has the design that it does. Honeybee research has historically been concentrated in Europe and North America, and so has inevitably emphasized just one species of honeybee, *Apis mellifera*, and the way it lives in the northern latitudes of these regions. Unless stated otherwise, the discussion can be assumed to refer to these studies of the temperate-zone races of *A. mellifera*. However, in the final chapter, I will emphasize the ecology of the other species in the genus *Apis* and the races of *A. mellifera* having non-European origins. Though the mechanisms of honeybee social life will not be the prime subject of this book, they will be discussed frequently, since understanding the minute operational details of an adaptation often casts light on underlying selective pressures. Moreover, knowledge of the machinery of an animal's behavior provides behavioral ecologists with ways to probe the adaptive significance of the behavior experimentally. Reciprocally, the ecological view illuminates the path to understanding the mechanisms of social life. By studying how an animal lives in its natural environment, a biologist gains a clear picture of its full behavioral repertoire, and develops a heightened intuition for the physiological processes which underlie its behavioral adaptations. One main theme of this book is, therefore, to exemplify the synergism which arises from a balanced combination of physiological and ecological studies of social behavior.

Individual-Level versus Colony-Level Selection

The logical first step toward understanding adaptation in honeybees is to identify the level (or levels) of biological organization at which natural selection operates in social insects. Since the founding of the theory of evolution by natural selection, most biologists interested in insect sociality have emphasized selection at the level of colonies (Darwin 1859, Weismann 1893, Wheeler 1911, Sturtevant 1938, Emerson 1960, Wilson 1971, Oster and Wilson 1978). According to this view, the morphology, physiology, and behavior of an individual social insect are adapted to benefit its colony's reproductive success, not necessarily its own. This is a group-selection view of evolution, but one which is at least plausible, given that social insect colonies are discrete groups and that they possess variation, heritability, and fitness differences—three properties an entity must possess if it is to evolve by natural selection (Lewontin 1970). Relatively recently, students of the social insects have emphasized colony-level selection less and have focused attention instead on individual-level (or even gene-level) selection (Hamilton 1964, 1972, Williams 1966a, Alexander 1974, West Eberhard 1975, Dawkins 1976, 1982). According to this view, each member of a social insect colony has been selected to maximize its own reproductive success (inclusive fitness), even if this creates inefficiency and reduces its colony's overall fitness. The impressive group behaviors of social insects, such as cooperative food collection and precise control of nest temperature are, according to this viewpoint, simply statistical summations of many individuals' ultimately selfish actions.

There can be little doubt that individual-level selection is important in social insect evolution and therefore that colony-level selection is not of universal importance. Proof of selection having operated on individuals comes from numerous reports, involving a wide array of species, of workers laying eggs (reviewed by Hamilton 1972, Oster and Wilson 1978) or of dominance interactions among colony members (reviewed by Wilson 1971; see also Cole 1981, Franks and Scovell 1983). Such behaviors certainly decrease efficiency at the colony level but make sense in terms of individuals competing for reproductive success.

In honeybees, there are two dramatic examples of conflict among individuals that leads to a decrease in colony efficiency. The most familiar is the fights to the death (by stinging) among newly emerged queens. The benefit to an individual queen of killing her rivals is clear: undivided motherhood of the colony's next crop of reproductives. However, this combat between queens may work to the detriment of the colony if all the queens kill each other, or if the lone survivor is later preyed upon when she ventures outside the nest for mating. A second, even clearer product of individual-level selection is the behavior of workers in a colony that loses its queen and fails to rear a replacement. The best thing that the workers can do in this situation in order to maximize colony fitness is to produce one final crop of male reproductives reared from the unfertilized eggs which workers can lay. But rather than rear these males as cooperatively and efficiently as possible, disharmony erupts in the nest as workers compete to provide the eggs that will produce the males. Workers with active ovaries are mauled by workers with inactive ovaries (Sakagami 1954, Korst and Velthuis 1982). Frequently a half a dozen or more eggs will be laid in each cell in the broodnest, in stark contrast to the orderly one-egg-per-cell pattern when a queen is present and despite the fact that only one drone at a time can be reared in a cell (Fig. 1.1). A third possible product of individual selection among honeybees is the production by some colonies of several small reproductive swarms following the large primary swarm (Allen 1956, Winston 1980). Although no data are available to prove the point, it seems likely that these ''afterswarms'' are detrimental

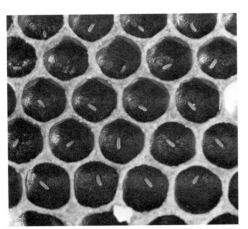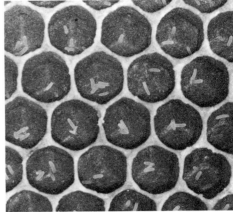

Figure 1.1 Comparison of egg laying patterns in queenright (left) and queenless (right) colonies. Queen-laid eggs are neatly deposited one to a cell, an arrangement which fosters efficiency in brood rearing. In the absence of a queen, a less efficient pattern of multiple eggs per cell appears as the workers compete among themselves for reproductive success.

to the parent colony, both because they further deplete the parent colony's worker force and because they are so small that they probably have little chance of surviving to maturity. On the other hand, the queen departing with an afterswarm may stand a better chance of survival than if she had remained to fight for control of the parent colony, and the workers closely related to her may achieve higher reproductive success by leaving with her to found a new daughter colony, as compared with staying with the parent colony and helping a more distantly related queen (Getz et al. 1982).

Clearly individuals are an important unit of selection in social insects, even in the honeybee with its complex colonial organization, but this by no means precludes the importance of colony-level selection. Just how potent is colony-level selection? Unfortunately, this question is not readily answered, in large measure because individual-level and colony-level selection should frequently promote the same pattern of adaptation. For example, in a queenright honeybee colony (one in which the workers are not laying eggs and thus achieve reproductive success indirectly via their mother queen [see Chapter 3]), both an individual worker's inclusive fitness as well as her colony's fitness are promoted by the worker performing such tasks as brood rearing, comb construction, and food collection as efficiently as possible. The larger the pool of resources assembled by the colony and the higher the efficiency of their use, the greater the number of reproductives the colony can manufacture and the higher each worker's inclusive fitness. In fact, it may be precisely because selection at individual and colony levels can operate in concert that certain species, such as the honeybee, possess such elaborate social organization. It might seem that a colony member working so as to decrease her inclusive fitness but increase colony fitness would prove the superior importance of colony-level selection. One could argue, for example, that this explains why workers refrain from rearing sons and instead help their mother rear their brothers even though they are more closely related to their sons than to their brothers (see Chapter 3). However, such apparently altruistic behavior can also be explained by adaptation at the individual level. One possible explanation is that the seemingly selfless worker simply has been manipulated by another individual seeking to boost its own inclusive fitness. The phenomenon of worker bees not laying eggs when a queen is in the nest could reflect precisely this process, with the queen perhaps dominating the reproductive activities of the workers via her inhibitory, "queen-substance" pheromone. In summary, I know of no observation on honeybee biology which unequivocally demonstrates the action of colony-level selection working at the expense of individual interests.

Our fascination with a colony of honeybees, army ants, or other advanced social insects is born largely out of curiosity about its overall achievements as an animal group. The intricate internal organization of a colony's foraging behavior, the efficiency expressed in nest design, the high precision of nest temperature control, all suggest functional, adapted organization of the colony as a whole. This intuitive feeling may be correct, but given that strong evidence exists only for selection at the level of individual social insects, it seems correct for now to explore adaptation in honeybees as far as possible in terms of individual-level selection, but also to keep in mind the possible role of colony-level selection, especially wherever individual and colony interests coincide.

2 Honeybees in Nature

Fossil Honeybees

The origins and evolution of the genus *Apis* can be discussed with fair confidence, a consequence of the outstanding richness of the honeybee's fossil record. Many fossils are so beautifully preserved in amber or shale that they can be examined as modern specimens (Zeuner and Manning 1976, Culliney 1983). The earliest bees in the genus *Apis* are known from fossils uncovered in Germany and France dating from the early Oligocene, about 35 million years ago (Fig. 2.1). Whether these earliest *Apis* formed societies or lived solitarily is a mystery. However, some signs point to sociality in *A. armbrusteri*, a honeybee of the late Oligocene or early Miocene. Specimens of this species closely resemble workers in a living honeybee species, *A. mel-*

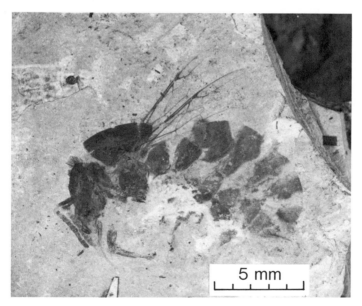

Figure 2.1   Photograph of the fossil honeybee, *Apis henshawi*, Cockerell from the Oligocene of West Germany. (Courtesy of Laurie Burnham.)

lifera, in both overall body form as well as in such detailed morphological characters as wing shape, wing venation, and hind leg design. Also, one fossil of *A. armbrusteri* contains 17 bees closely packed in a small rock, perhaps a fragment of a fossilized swarm.

Among the oldest fossil bees, found in Baltic Amber of late Eocene age (about 40 million years old, Burleigh and Whalley 1983), are specimens which resemble honeybees and are thought to have given rise to the genus *Apis*. Because they differ significantly in thorax shape, leg structure, and wing venation from *Apis* specimens, these older, extinct bees have been classified in a distinct genus, *Electrapis* (= amber honeybees). Assuming that the genus *Apis* arose from an *Electrapis* species, and given that all fossil and living *Apis* closely resemble each other, it appears that the evolutionary history of honeybees comprises an initial period of rapid morphological change which lasted some 10 million years, from the late Eocene to the late Oligocene, followed by an approximately 30 million year period of relative stasis in morphological evolution, from the Miocene to the present. This pattern suggests, assuming that honeybee social behavior and worker morphology have evolved in tandem, that the social organization found today in the genus *Apis* has a history of some 30 million years.

Systematics of Living Honeybees

Today there are probably just four or five, but possibly six or more living species of *Apis*. This genus of bees is native to Europe, Africa, and Asia, including such continental islands as Japan, Taiwan, the Philippines, and most members of the Indonesian archipelago. Honeybees also thrive in North America, South America, and Australia, but only since European man introduced them at various times during the seventeenth to nineteenth centuries. The biogeography and systematics of the genus *Apis* have been reviewed most recently by Maa (1953), Ruttner (1968a, 1968b), and Ruttner, Tassencourt, and Louveaux (1978).

The most widely known *Apis* species is the western hive bee, *Apis mellifera* (Fig. 2.2). Its native distribution extends from the steppes of western Asia, through Europe as far north as southern Norway, and into all of Africa except for its great desert areas. This is the species which Europeans spread over the world for use in apiculture. Wherever *A. mellifera* exists today, humans exploit it as a honey and wax producer and as a pollinator. *Apis cerana*, the eastern hive bee, closely resembles *A. mellifera* morphologically and in nesting behavior, but is clearly a distinct species (Fig. 2.3). When brought together, these two normally allopatric species never hybridize (Sharma 1960, Ruttner and Maul 1983). The drones of each species are attracted to queens

Figure 2.2   A worker honeybee, *Apis mellifera*, foraging on buckwheat flowers. Note the load of pollen packed on the hind leg.

of the other species (Ruttner and Kaissling 1968), but if mating occurs, physiological defects in the blastodermal stages of development prevent the formation of hybrid embryos (Maul 1969). Two other species of *Apis*, the dwarf honeybee, *A. florea* (worker length, 7–8 mm), and the giant honeybee, *A. dorsata* (worker length, 16–18 mm), both differ strikingly from *A. mellifera* and *A. cerana* in nesting habits and in geographical distribution. Whereas *A. mellifera* and *A. cerana* both nest inside protective cavities such as small caves or tree hollows, a behavioral trait which has helped both species to penetrate into cold temperate regions, a colony of *A. florea* or *A. dorsata* constructs its nest in the open, hanging beneath a tree branch or an overhanging rock (*A. dorsata* only) (Fig. 2.4). Evidently the high exposure of their nests restricts these bees to warm southern Asia, including islands such as Sri Lanka, the Philippines, and all but easternmost Indonesia (*A. dorsata* reaches Timor). *Apis florea* ranges west to Oman; the western limit of *A. dorsata* probably lies in Pakistan.

Each of the taxa now recognized as *A. florea* and *A. dorsata* may actually include two or more species. Michener (1974), for example, reports seeing male specimens of the *A. florea* group which seemed to belong to two different species (*A. florea, A. andreniformis*). Similarly, morphological comparison between highland and lowland specimens of workers of the *A. dorsata* group suggest that the highland bees constitute a distinct species, *A. laboriosa*, which inhabits the mountainous areas of Nepal, northeastern India, and southern China, at elevations mainly between 1000 and 3000 meters (Sakagami et

Apis florea

Apis cerana

Apis dorsata

10 mm

Figure 2.3   Workers of three species of honeybee native to southern Asia.

al. 1980, Roubik et al. 1985). Workers of *A. laboriosa* are distinguished from those of *A. dorsata* by several traits. For example, the thorax is 11 percent wider, the thoracic hairs are 33 percent longer, and, especially, the ocellar area is conspicuously flatter. Conclusions about the validity of *A. andreniformis* and *A. laboriosa* as distinct species, however, must await careful analyses of specimens collected from the zones where the ranges of the proposed species overlap. For *A. laboriosa* and *A. dorsata*, one such zone is the area lying between 1100 and 1400 meters elevation on the southern slope of the central Himalayas in Nepal.

Each species of *Apis* varies strikingly over its range, exhibiting geographical races which differ not only in body color and structural details—such as

Figure 2.4   Nests of four species of honeybee: (A) *Apis mellifera*; (B) *A. florea*; (C) *A cerana*; (D) *A. dorsata*. The nests of the two cavity-nesting species, *A. mellifera* and *A cerana*, were exposed by cutting open their tree-cavity nest sites. In both of these nests the bees have been killed and most have dropped off the combs.

tongue length, body size, and hair coverage—but also in social behavio (reviewed by Ruttner 1975a, 1975b). Presumably these differences reflec adaptation by each local population to the particular conditions under whicl it lives. Especially clear examples of ecologically significant differences ii behavior have emerged from studies of two European races of *A. mellifera A. m. ligustica* (Italian bees) and *A. m. carnica* (Carniolan bees). As explaine(by Ruttner (1968b), these two races differentiated during the last Ice Ag when honeybees in Europe existed as isolated populations in a few southerl refuge zones. Two of these zones, the Apennine and Balkan peninsulas, wer the sources of the Italian and Carniolan races, respectively. When the glacier retreated ten thousand years ago, *A. m. carnica* expanded northward into the entire Danube Valley (Yugoslavia, Hungary, Romania, Bulgaria) and onto the western steppes of Asia, but *A. m. ligustica* was confined to Italy by the Alps. Experimental studies (Boch 1957, Jaycox and Parise 1980, 1981, Gould 1982) have revealed that Carniolan bees, relative to Italian bees, range farther afield when foraging, choose larger nest cavities, and disperse farther from the parent nest when reproducing their colonies. These interracial behavioral differences may be rooted in geographical differences in overwintering conditions. Carniolan colonies, native to a part of Europe possessing a continental climate with long, severe winters, may require more stored food to survive winter (hence greater storage space and foraging on a broader scale) than do Italian colonies, which are adapted to a Mediterranean climate with short, mild winters. Other well-documented behavioral differences among various honeybee races include aggressiveness (Stort 1974, Collins et al. 1982), attraction to particular floral scents (Koltermann 1973), annual rhythm of brood production (Adam 1968, Ruttner 1975a), age-labor schedule (Winston and Katz 1982), and skill in learning visual cues near food sources (Menzel et al. 1973, Hoefer and Lindauer 1975). This wealth of interracial differences, each of which may possess strong ecological significance, underscores the importance in studies of honeybee ecology of working with racially pure, or at least genetically well-defined, colonies of bees.

"Domestication" of Honeybees

Man's association with the honeybees extends back into prehistory—perhaps even to times before our ancestors were humans. Although most of the details of this history can never be known, studies in archaeology and anthropology (reviewed by Townsend and Crane 1973, Crane 1975, 1983) suggest that man's methods for exploiting honeybees have developed repeatedly in a consistent pattern among the various peoples of the Old World. The initial phase probably consisted simply of hunting for wild colonies of bees. At this point, man differed little from other mammals, though the use of smoke to subdue bees and probably also containers to hold honeycombs perhaps made early man an especially efficient predator on honeybees. This first stage of honeybee exploitation is vividly documented by mesolithic rock paintings in South Africa, India, and Spain, and by accounts of the honey-gathering techniques of primitive peoples in Asia and Africa. An intermediate stage, between bee hunting and true beekeeping, involved ownership of scattered wild colonies from which honey was seasonally harvested. In northern Europe, whose forests were once rich with honey, this second phase of honeybee exploitation occurred in the Middle Ages. Documents still exist that illustrate in detail the extensive tree beekeeping in the forests surrounding Moscow before the 1700's (Galton 1971). Records from the Morozov Estates for 1667, for example, describe 8 bee forests which averaged 95 km² in area and contained about 0.4 honeybee colonies/km². These forest colonies were exploited by bort-niks—beemen—who discovered trees occupied by bees and marked them with ownership signs, and then harvested honey and wax in the late summer by scaling each tree, chiseling out an opening beside the bees' nest, and slicing away some portion of the honey-filled combs.

True beekeeping began when man created apiaries by transporting the nests of wild colonies to sites near human dwellings, and, somewhat later, by constructing containers specifically for bees: hives. Only the two cavity-nesting species, *Apis mellifera* and *A. cerana*, figure strongly in these last two stages of honeybee exploitation. In many parts of the world, including Europe, these stages were reached only in the last few centuries. Presumably the earliest hives consisted simply of the tree section containing a natural nest. Indeed, the German word for beehive—*Bienenstock* (*Stock* = tree-trunk)—is one indication of the modernity of keeping bees in hives. Subse-quent hives were equally simple man-made containers—inverted clay pots or straw baskets—built of cheap materials. One major shortcoming of these early hives was that bees anchored their combs to the roof and sides of the container, thus making it impossible to inspect or manipulate colonies in a routine manner. It was not until 1851, scarcely more than a century ago, that Lorenzo L. Langstroth invented a hive in which the combs were freely removable for inspection or insertion in another hive (Naile 1976), and thereby established the foundation for modern beekeeping and honeybee research.

Unlike most other organisms exploited by man for thousands of years, the honeybee has not been strongly modified through artificial selection (Roth-enbuhler 1958, 1982, Ruttner 1968c). It is only within the last 100 years or so that we have assembled the knowledge needed to breed bees at all, and only within the last 35 years that we have mastered the skills which allow truly efficient artificial selection. Understanding this unusual state of affairs requires a brief review of the history of bee breeding. First, the basic facts of honeybee reproduction were not understood until the mid 1800's. Although Swammerdam had determined by 1680 that queen bees are females and lay their colonies' eggs, and by 1771 Janscha had discovered that each queen flies out of her nest to mate, it was not until 1848 that Dzierzon revealed that males and females arise from unfertilized and fertilized eggs, respectively, and so finished sketching in the broad picture of honeybee reproduction.

The next milestone in the history of honeybee breeding was the development between 1850 and 1890 of methods for rearing numerous queens from the brood of one selected colony (Pellett 1938). This advance, one product of Langstroth's movable-frame hive, made planned bee breeding possible, but still in only a crude form. The primary residual problem was that bee breeders could not control which males mated with the queens they had selected, since queens only mate in free flight. A further difficulty arises from the queen honeybee's practice of mating with several drones, thus giving each colony a complex genetic composition (see Chapter 3). This confuses the task of evaluating each queen based on the properties of her colony. It was not until the 1940's, when reliable methods of instrumentally inseminating queens were perfected, that complete control of honeybee parentage became possible, permitting truly effective artificial selection of honeybees (Watson 1928, Laidlaw 1944, Mackensen and Roberts 1948).

Before the advent of instrumental insemination, bee breeders relied either on selection among females only, an inherently inefficient technique, or on isolated mating stations (supposedly containing only males of selected mothers), a laborious and fallible approach to controlling matings. Although beekeepers believe that the first technique has improved their bees—rendering them gentler, harder working, and less inclined to swarm—documentation of whatever progress was made is lacking (Rothenbuhler 1958). In contrast, the accomplishments ascribed to the use of isolated mating yards are few in number, but thoroughly reported. The most striking achievement was a strain of bees highly resistant to a larval disease called American foulbrood (causative agent: *Bacillus larvae*). Colonies were inoculated with pieces of comb containing 75 larvae which died from the disease, then virgin queens and drones reared by the disease-resistant colonies were taken to an isolated mating station in the middle of a 100 square kilometer citrus orchard. Fifteen years of testing and breeding reduced the fraction of inoculated colonies which acquired the disease from 72 to 2 percent (Park 1937, 1953, Park et al. 1937).

The introduction of instrumental insemination set the stage for rapid progress in bee breeding, and through its use several novel strains of honeybee have been developed. These include bees with a strong preference for alfalfa pollen (Nye and Mackensen 1968), bees which collect sugar solution at especially high or low rates (Rothenbuhler et al. 1980), bees with intense brood production and high honey yield (Cale and Gowen 1956), and bees resistant to hairless-black syndrome, a viral disease of adult honeybees (Kulinčević and Rothenbuhler 1975). However, these specially selected lines of bees have rarely become incorporated into the stocks used by commercial breeders of queen bees. Possible exceptions to this rule are certain disease-resistant and vigorous honey-producing strains. In general, man is just beginning to modify honeybees through artificial selection. This is of major importance to biologists interested in their natural history and ecology, because it means that honeybees remain primarily the product of natural, rather than artificial, selection.

Without doubt, man's principal influence on the genetic composition of honeybees has occurred through his mixing of the various geographical races of honeybees (Rothenbuhler 1958, Adam 1968). The bees in North America, for example, are an amalgamation of genetic types introduced from the 1600's to the early 1900's, most importantly the four European races, *Apis mellifera mellifera, A. m. ligustica, A. m. caucasica,* and *A. m. carnica,* with *A. m. ligustica,* the Italian bee, predominant by far (Phillips 1915, Pellett 1938). This complicates ecological studies because colonies will not necessarily possess an integrated set of co-adapted traits, but may instead exhibit some odd combination of the properties of hybridized races. Probably the best solution to this problem is to work only with racially pure colonies, carefully recording the type of colonies used by depositing voucher specimens in an established insect collection. An alternative approach is to conduct ecological studies of feral bees inhabiting areas lacking intensive beekeeping, and hence not subject to the strong inflow of genes that results when beekeepers introduce queens purchased from geographically distant queen-breeding companies. Within such an area, the wild colonies are probably well adapted to local conditions, having experienced many years of strong natural selection. A region that satisfies this condition is found in the vicinity of Ithaca, New York, where I have conducted several ecological studies on honeybees (Seeley 1978, Seeley and Morse 1976, 1978, Visscher and Seeley 1982). In the forests here wild colonies occur at a density of approximately 1 colony/km²; a small number of beekeepers' colonies also exist on the edges of these forests. The wild colonies are primarily descendants of Italian bees introduced by bee-keepers since the late 1800's. Though these bees were originally adapted to a mild Mediterranean climate, they now thrive in a region characterized by short, hot summers and long, harsh winters, suggesting that they have adapted through natural selection to the local climate.

The Importance of Studying Honeybees in Nature

Ecological studies of honeybee behavior must be conducted with bees living under natural conditions. This methodological principle takes into account the molding of honeybee behavior by natural selection, and the often re-markably precise fit between a bee's behavior and its external environment. When displaced from a natural setting, bees are frequently deprived of critical stimuli—the scents of flowers or pheromones, the seasonal rhythm of changes in daylength, the dark forms of approaching predators—to which their be-havior is tuned. Moreover, they are often bombarded with abnormal stimuli—sharp vibrations of the colony, bright illumination of the nest, superabundance of food. The frequent result is inappropriate behavior, a truncated behavioral repertoire, or some other failure of a bee's normally well-adapted actions.

A striking example of honeybees behaving maladaptively—though to man's benefit—when in an unnatural environment is the tendency of honeybee colonies in beekeepers' hives to refrain from colony reproduction and instead to stockpile several times as much honey as they need for winter survival. Thus, whereas a typical wild colony reproduces (swarms) at least once each year but accumulates only about 20 kilograms of honey over a summer (Seeley and Morse 1976, Seeley 1978, Winston 1980), only about 25 percent of the colonies in an apiary swarm each year and their annual honey production is often 50 to 100 kilograms per hive (Simpson 1957a, Allen 1965a). This shunting of resources into storage and away from reproduction benefits beekeepers, whose income reflects their honey harvest, but hurts the bees, whose genetical fitness would be greater if they concentrated more on reproduction and less on honey storage. The critical step taken by beekeepers to disrupt the honeybee's life history is apparently the provision of excess nest space. The nests of wild colonies in the northeastern United States typically fill only 30 to 80 liters of space, while beekeepers' hives generally provide 125 to 250 liters of nest space (Seeley and Morse 1976, Seeley 1977). The superabundant volume of beekeepers' hives evidently inhibits colonies from swarming by eliminating a major stimulus for swarming: overcrowding of bees inside the nest (Simpson 1958, 1973, Winston et al. 1980, Lensky and Slabezki 1981). Also, the unusually large amount of empty comb in beekeepers' hives probably creates an abnormally powerful stimulus for bees to amass honey stores (Rinderer and Baxter 1978, Rinderer 1982).

Besides running the risk of observing sociological oddities, the biologist studying bees in a disturbed environment is also liable to overlook the ecological significance of any behaviors which are performed correctly. For example, "planing" is a behavior in which ten or more bees rock forward and backward with their heads down, front legs bent, while scraping the wooden surfaces around the hive entrance with their mandibles (Gary 1975). The adaptive significance of this planing behavior seems obscure if one observes it only on hives constructed of smooth lumber. In contrast, when one sees bees inhabiting a tree using this behavior to smooth the rough bark around their nest's entrance, its adaptiveness seems clear. A smooth entranceway facilitates the heavy traffic of foragers scrambling in and out of the nest and probably also simplifies the surveillance by guard bees against small intruders, such as ants.

A second example of confusion about functional design concerns the honeybee's dance-language system of recruitment communication. Using this system, bees clearly have the ability to share information about food sources 10 or more kilometers from the nest, since bees will recruit nestmates to a dish of sugar syrup located 10 kilometers from their hive (von Frisch 1967). Early studies of the foraging range of colonies, however, found that 80–90 percent of a colony's foragers worked less than 1.5 km from their nest, and that none foraged beyond about 3 km (Knaffl 1953, Beutler 1954, Levin and Glowska-Konopacka 1963). Why should bees have the capacity to communicate about food sources three times farther from the nest than they ever travel? In fact, this puzzle was simply an artifact of unnatural study conditions; the colonies studied were either unusually small, or lived amidst vast tracts of forage created by agriculture. A study of a full-size colony living in a temperate deciduous forest revealed that honeybees in nature routinely forage 6 km and occasionally up to 10 km from their nest (Visscher and Seeley 1982). Thus this real-world view of honeybee foraging demonstrated a close match in spatial scale between this bee's foraging operations and its communication system. In general, studies of honeybees living in nature are essential if we are to view their normal behavior and understand the evolutionary forces by which this behavior has been molded.

3

The Honeybee Societies

Kin Structure of Colonies

The fundamental social structure of a honeybee colony is that of a matriarchal family. At the heart of each colony lies one long-lived female, the queen, who is the mother of the thirty thousand or so members of a typical colony. Approximately 95 percent of each queen's offspring are workers, daughters which never mate and never lay any eggs so long as their mother is alive (Jay 1968, 1970). Instead, they help their mother survive and reproduce, performing all tasks in the colony except the production of eggs. The other 5 percent of a queen's offspring develop into sexual reproductives—queens and drones. Queens control the sex of their offspring by a most elementary mechanism: males arise from unfertilized eggs and therefore are haploid, whereas females arise from fertilized eggs and are diploid (Kerr 1969, Michener 1974, Crozier 1977). Whether a fertilized egg develops into a worker or queen depends on the composition of the food given the developing bee during the first three days of larval life. The critical difference between the food fed to queen and worker larvae is apparently simply the concentration of hexose sugars; royal jelly and worker jelly contain about 35 percent and 10 percent sugar, respectively. Larvae reared on worker jelly experimentally fortified with glucose and fructose (200 mg of each per gram of worker jelly) produce queens. Evidently the sweetness difference triggers different larval feeding rates, different levels of juvenile hormone during development, and ultimately different developmental programs for the two types of female bees (Beetsma 1979, de Wilde and Beetsma 1982).

Although all members of a honeybee colony share the same mother, the female members do not all share the same father, a fact of major importance in understanding the evolution of honeybee social life. Mating by queen honeybees occurs only during the two-week period immediately following each queen's emergence as an adult. During this time, a queen makes several flights from her nest, receiving sperm from males on one to four of these flights. By the close of her mating period, each queen has stored about 5 million sperm in her spermatheca, a sufficient supply for her potential lifespan of about three years (Roberts 1944, Ruttner 1956, Woyke 1960, 1964). The average number of males inseminating a queen has been estimated in several ways, including comparing the sperm counts of sexually mature males with the number of sperm in recently mated queens, and applying probability models for the appearance of genetic markers in a queen's offspring. The estimates range from 7–17 males per queen, with intermediate values of 10–12 males per queen probably most accurate (Taber and Wendel 1958, Woyke 1960, Kerr et al. 1962, Adams et al. 1977). Although the precise pattern of a queen's use of the different males' sperm remains unknown, several studies involving either allozymes or phenotypic markers reveal that at any one time naturally mated queens draw upon the sperm of at least two or three males, and probably many more (Taber 1955, Kerr et al. 1980, Page and Metcalf 1982, Page et al. 1984). However, the sperm of different males is used nonrandomly, perhaps as a consequence of incomplete mixing of sperm stored in a queen's spermatheca. Thus, for example, the percentage of bees reared with a certain phenotypic marker (light-colored chitin; cordovan strain) in one of Taber's (1955) study colonies (containing a naturally mated queen) rose from 7 to 45 percent over a 90-day period. Random sampling of the sperm would have produced small fluctuations around an average value of about 22 percent cordovan bees.

The advantage to individual queens of multiple insemination evidently lies in ensuring high brood viability, and ultimately high queen fitness, despite the large genetic load associated with the honeybee's system of sex determination (Page 1980). Sex in honeybees is ultimately determined by a single locus with multiple alleles. Heterozygotes at this locus become females while homozygotes or hemizygotes (haploid individuals) become males. However, the homozygotes (diploid males) are killed by workers early in larval development (Woyke 1963). This prevents waste of food and space in the nest, since diploid males, whose average testes volume is only about 10 percent that of testes in a haploid male, are nearly sterile (Woyke 1973a). By drawing upon the sperm of several males, rather than just one, or two, a queen boosts the probability that her diploid brood's viability is near the population average of $[1 - (1/K)]$, where K represents the number of sex alleles in the population. Estimates of the number of these alleles range from 6 to 19, depending upon the population (Laidlaw et al. 1956, Kerr 1967, Woyke 1976, Adams et al. 1977).

The net effect of the queen's promiscuity and the mixing of different males' sperm is the complexity of genetic relationships within a honeybee colony shown schematically in Figure 3.1. The most important point here is that the females (workers and queens) produced in a colony are not all full sisters. Instead, they constitute several patrilineal groups, with females in the same group related as full sisters ($r = 0.75$) and females in different groups related as half sisters ($r = 0.25$).

A.

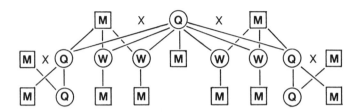

B.

RELATIONSHIP (X, Y)	$r_{X, Y}$	RELATIONSHIP (X, Y)	$r_{X, Y}$
QUEEN, DAUGHTER	0.50	WORKER, HALF SISTER	0.25
QUEEN, SON	0.50	WORKER, BROTHER	0.25
QUEEN, GRANDSON OR GRANDDAUGHTER	0.25	WORKER, SON	0.50
QUEEN, FULL SISTER	0.75	WORKER, FULL NIECE OR NEPHEW	0.375
QUEEN, HALF SISTER	0.25	WORKER, HALF NIECE OR NEPHEW	0.125
WORKER, FULL SISTER	0.75		

Figure 3.1 Genetic relationships between members of a honeybee colony. (A) Schematic genealogy showing three generations: (1) the mother queen and her multiple mates, (2) the daughters (workers and queens) and sons of the mother queen, and (3) the reproductive grandchildren (queens and males) of the mother queen. Circles represent females (diploid) and squares represent males (haploid). (B) Coefficients of relatedness for the important relationships arising in honeybee colonies, calculated assuming there is no inbreeding (a valid assumption for honeybees; see Chapter 5). Relatedness $r_{X,Y}$ between two individuals X and Y denotes the probability that Y contains a gene identical by descent with a random gene at the same locus, sampled from X.

Queens and Workers: Reproductive Division of Labor

Although nearly all the thirty thousand or so bees in a honeybee colony are females, only one—the queen—normally lays any eggs. This sharp division between reproductive (the queen) and nonreproductive (the workers) females is documented by several lines of evidence. For one, dissections of thousands of workers from colonies containing queens have consistently shown that 90 percent or more of the workers possess completely inactive ovaries and the remaining 10 percent or so possess ovaries whose ovarioles are slightly swollen but only contain eggs in the most rudimentary stages of development (Verheijen-Voogd 1959, Jay 1968, 1970, Jay and Jay 1976, Kropáčová and

Haslbachová 1969) (Fig. 3.2). A second indicator of worker sterility in queen-right colonies is the effectiveness of a piece of beekeeping equipment called a "queen excluder." In essence, a queen excluder is a screen whose apertures allow the passage of workers but not of the somewhat larger queens (Dadant 1975). It allows a beekeeper to confine a colony's queen to a certain region of the hive and thereby segregate the colony's brood and honey stores. Brood never appears in the regions of hives from which queens have been excluded, providing strong evidence that workers with a queen refrain from laying eggs.

Workers possess the physiological apparatus for producing viable eggs, but it is only activated when a colony becomes hopelessly queenless. For example, if a colony's queen is killed and all the replacement queens subsequently reared by the workers from the last of their mother queen's eggs are also destroyed, thereby dooming the colony, the ovaries of the colony's workers will gradually enlarge. Fifteen days after the loss of the original queen, only about 5 percent of the workers possess ovaries containing mature eggs, but

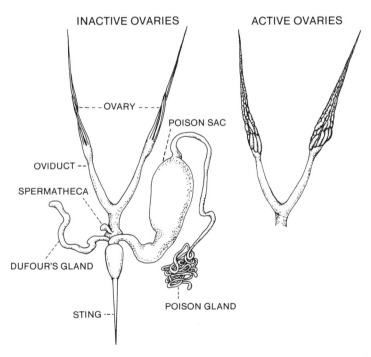

INACTIVE OVARIES   ACTIVE OVARIES

- - - - OVARY - - - -

POISON SAC

OVIDUCT - -

SPERMATHECA

DUFOUR'S GLAND

STING - -

POISON GLAND

Figure 3.2   Left: ovaries and associated organs of a worker honeybee possessing inactive ovaries. (After Snodgrass 1956, © Cornell University; used by permission of Cornell University Press.) Right: oviducts and active ovaries of a queenless worker showing the development of eggs. (After Michener 1974.)

after 30 days this proportion has expanded to about 50 percent (Jay 1968, 1970, Sakagami 1954). The extent of worker ovary activation depends on many factors, with season and colony nutritional status two of the more important influences. Queen loss in early summer and abundance of stored pollen both favor ovary development in queenless workers (Velthuis 1970).

The queen's presence is continuously signaled to the workers in a colony by a pheromone, (E)-9-oxo-2-decenoic acid (9-ODA), which is secreted by the queen's mandibular glands. Although this fatty acid is sufficiently volatile to function as the honeybee's sex attractant pheromone (Gary 1962) and as an indicator of the queen's presence in an airborne swarm of bees (Avitabile et al. 1975), inside a nest the 9-ODA signal of queen presence is transmitted via contacts between individuals. When a queen pauses between bouts of egg laying, many of the workers nearby lick her or brush their antennae over her, thus adsorbing trace amounts of 9-ODA (Fig. 3.3). Next they scurry through the nest as messengers, contacting other workers and so disseminating the message "the colony contains a queen" (Velthuis 1972, Seeley 1979, Ferguson and Free 1980). The presence of brood also inhibits the development of workers' ovaries, and possibly provides an inhibitory signal even stronger

Figure 3.3 Queen bee surrounded by workers. The bees touching the queen with their antennae are wiping trace amounts of (E)-9-oxo-2-decenoic acid off the queen. By traveling about the nest and contacting other workers, they help disperse this chemical signal of the queen's presence.

than the queen's pheromone (Jay 1970, Jay and Jay 1976, Kropáčová and Haslbachová 1970, 1971). Thus when a colony lacks an adult queen temporarily, for example during colony reproduction (see below), the yet unemerged brood of the recently departed queen keeps the workers' ovaries inactive until the new queen emerges to head the colony.

Given that natural selection shapes individuals to be skilled at propagating their genes, the sterility of honeybee workers in the presence of a queen constitutes, at first glance, a profound puzzle. Why should a queen's worker-daughters forego their own reproduction and instead devote their entire lives to helping their mother produce offspring? Fortunately, enough is now known about the colony kin structure, population structure, and queen-worker relationship of honeybees in particular, and about the evolution of altruism in general, to enable us to begin piecing together a solution to the puzzle of worker sterility in honeybees.

Central to this explanation are the ideas of kin selection, first discussed formally by Hamilton (1964, 1972), and later refined for the social insects by West Eberhard (1975, 1978), Trivers and Hare (1976), Craig (1979, 1983) and others (reviewed by Oster and Wilson 1978, Crozier 1979, Starr 1979). The fundamental insight of kin selection theory is that "altruism" is genetically advantageous to an individual (Ego) if the following condition holds:

$$B/C > r_{A_Y}/r_{B_Y}, \tag{3.1}$$

where B and C represent respectively the benefit to beneficiary and cost to the altruist (both measured in number of reproductively successful offspring), and r_{A_Y} and r_{B_Y} denote respectively the probabilities of Ego having a given gene in common by descent with the young of the altruist (A) or with the young of the beneficiary (B). (Ego can be the altruist, the beneficiary, or simply a relative of the altruist and beneficiary.) Is this condition fulfilled for queen and worker honeybees?

In honeybees, the queen is the beneficiary of altruistic acts performed by her worker-daughters. Thus from the queen's perspective, $r_{A_Y} = 0.25$ and $r_{B_Y} = 0.50$, and so the threshold benefit: cost ratio is $(0.25/0.50) = 0.50$. Altruism by a queen's daughters is advantageous to the queen so long as her daughters are at least 50 percent as efficient helping their mother produce offspring (reproductives) as they are in producing their own offspring. For reasons that will be discussed below, this is almost certainly true.

What is the threshold benefit:cost ratio for worker altruism from the perspective of a worker bee? Here Ego is the altruist, so r_{A_Y} is simply 0.50. The precise value of r_{B_Y}, the average relatedness of a worker to her mother-queen's reproductive offspring, is less obvious, for it depends on how many males inseminated the queen, the distribution of paternities among these males, and the sex ratio of the reproductive offspring. In the absence of inbreeding (a reasonable assumption for honeybees, given their mating system, described in Chapter 5), the average relatedness of a worker to female offspring of her mother (r_{WF}) can be calculated as:

$$r_{WF} = \frac{1}{2}\left(\frac{1}{2} + \sum_{i=1}^{n} f_i^2\right), \tag{3.2}$$

when the queen mates with n males and they are respectively responsible for proportions $f_1, f_2, \ldots f_n$ of her female progeny (Hamilton 1964). If all males contribute equally, this simplifies to:

$$r_{WF} = \frac{1}{2}\left(\frac{1}{2} + \frac{1}{n}\right), \tag{3.3}$$

which is the *lowest* possibile value of r_{WF} for a given value of n. Because males arise from unfertilized eggs, the relatedness of a worker to her mother's male offspring is independent of the paternity pattern in the colony. It is always 0.25.

The last parameter needed to calculate r_{BY} for workers is the ratio of investment between female (queens) and male offspring. Whether this is controlled by the queen or, as seems more likely in most social insects, by the workers (Trivers and Hare 1976) probably makes little difference in honeybees. On the one hand, the queen, related equally to her sons and daughters, would benefit most by equal investment in the two sexes. On the other hand, the workers have probably been selected to favor investing the fraction $(n + 2)/(2n + 2)$ of the reproductive effort in females, and $n/(2n + 2)$ of this effort in males, where n represents the number of males inseminating the mother queen (Charnov 1978a). (This assumes that all n males contribute equally to the female offspring.) Thus if 10 males inseminate the queen, the workers are expected to prefer 55 and 45 percent investment in female and male reproductives, respectively, values not greatly different from the queen's preferences of 50 percent for each sex. The average relatedness of a worker to the reproductive offspring of her mother-queen can now be estimated by:

$$r_{BY} = \frac{1}{2}\left(\frac{1}{2} + \frac{1}{n}\right)\cdot\left(\frac{n + 2}{2n + 2}\right) + \frac{1}{4}\left(\frac{n}{2n + 2}\right)$$
$$= \frac{n^2 + 2n + 2}{4(n + 1)n}. \tag{3.4}$$

Because this equation rests on the assumption that all n males inseminating a queen are equally represented in the female offspring, the estimates it provides are the lowest possible values of r_{BY} for a given n. Values for r_{BY}

as a function of n, calculated using equation 3.4, are shown in Figure 3.4. Assuming that queens mate with about 10 males, this figure shows that r_{B_Y} for a worker is no less than about 0.28. Therefore, the threshold benefit:cost ratio for worker altruism with their mother-queen the beneficiary is at most about $(0.50/0.28) = 1.8$. In short, it appears that for altruism by a worker honeybee to be favored by natural selection requires that a worker be nearly twice as efficient in rearing her mother's offspring relative to rearing her own.

Unfortunately, it is probably impossible to measure the benefit:cost ratio for worker honeybee altruism and thereby test these ideas empirically. Although the queen's benefit per worker helping is, in principle, measurable (in units of reproductive offspring gained), measuring the cost per worker of helping is a formidable challenge. The difficulty arises simply because workers never try to rear their own offspring as long as they can help their mother. However, given the honeybee's current social organization, it is easy to believe that a worker's aid to her mother yields at least 1.8 times as many offspring (reproductives) as would arise if the worker reproduced directly. A worker's egg-laying capacity of 50 per day is extremely modest compared to the queen's capacity of 2000 per day (Sakagami 1958, Ribbands 1953), but a more important limit to a worker's success in personal reproduction would un-

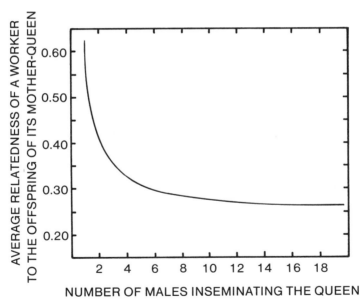

Figure 3.4 The relationship between number of males inseminating a colony's queen and the average genetic relatedness between the workers and reproductives produced in the colony.

doubtedly be her ability to care for her offspring. If she attempted to lay eggs in her mother's nest, her nestmates would probably maul her just as they maul laying workers in queenless colonies. (Most of any worker's nestmates are her half sisters; they are expected to prefer rearing their mother's sons ($r = 0.25$) rather than those of a half-sister ($r = 0.125$).) If the worker attempted to go it completely alone, she would face the many hurdles of solitary life, including constructing a nest, laying eggs, and feeding and guarding her brood. The behavioral programs of worker honeybees today, refined for social living over millions of years, are probably totally inappropriate for solitary life. Thus it seems likely that the cost of altruism by workers is negligible. At the same time, the benefit to a queen of a worker's aid is probably considerable. A worker's labor contributes directly to an established social group, where it can support operations such as colony defense, nest temperature control, and food collection, all of which are most efficiently performed by groups. Moreover, present-day honeybee queens and workers are well adapted to their respective roles of receiving and giving aid. Such attributes as the queen's immense egg-laying capacity, and a worker's communication skills and age-polyethism schedule (see below) only make sense in light of their complementary strategies of receiving and giving aid.

Of course, not all of a queen's daughters become workers; a tiny fraction form the next generation of queens. Unlike their worker-sisters, these queens show little if any altruism toward their mother. Several facts about honeybee biology indicate that the benefit:cost ratio of altruistic acts by a queen's queen-daughters falls far below the value experienced by her worker-daughters. First, the amount a queen can help her mother is probably minimal. Queens lack the various glands which produce wax, brood food, and alarm pheromones; their hind legs lack pollen-collecting hairs; their eyes, brains, and antennae are smaller than those of workers (Snodgrass 1956). Also, queens never participate in the ordinary duties of the hive such as cleaning cells, tending the young, or gathering food. After performing their nuptial flights, queen honeybees function as little more than egg-laying machines. The second indicator of a low benefit:cost ratio for a queen helping her mother and not reproducing is the high cost of such altruism. In contrast to a worker, a queen emerges into a world which, with few exceptions (such as rival sister-queens and predators encountered during mating flights), strongly fosters her reproduction. All queens surviving to head a colony are endowed with a helper force of several thousand workers. Moreover, many queen honeybees inherit their mother's completely constructed nest, already stocked with pollen and honey stores as well as developing bees, their mother having departed in a swarm a few days before her daughter-queens emerged as adults. In short, a queen would deliver scant benefits, but would incur massive costs, if she were to devote her life to helping her mother, and so there could be no selection for altruism by females that have developed as queens.

As we have seen, given the honeybee's current social organization, it is not terribly puzzling why worker-daughters find it genetically advantageous to be sterile and help their mother reproduce, and why queen-daughters find it genetically advantageous to seek personal reproduction. It remains a good deal more mysterious how worker-daughter sterility arose in the first place. Unfortunately, this question is rather intractable for honeybees. Not only can we not observe the past, but because there are no primitively social species of honeybees, we cannot even turn to comparative studies for insights into the origins of sterile helpers in this group of social bees. Perhaps at best we can simply delineate an array of possible scenarios.

To begin, one might argue that multiple insemination of females and mixing of the different males' sperm probably arose before the origin of reproductive division of labor in honeybees. This viewpoint follows from the consideration that a female can reduce the variance in brood viability by simultaneously using the sperm of several males, and that this advantage should apply to solitary as well as social female honeybees. Because multiple mating and sperm mixing reduce the average relatedness among sisters to well below 0.75, Hamilton's (1964) "3/4–relatedness hypothesis" (West Eberhard 1975) for why daughters in the social Hymenoptera stay and help their mothers probably does not apply for honeybees. By this hypothesis, if daughters can bias the sex investment ratio to favor females (Trivers and Hare 1976), then the higher average relatedness of a daughter to her siblings than to her offspring would favor the evolution of sterile daughters helping their mother reproduce.

An alternative explanation for the advent of sterile workers in honeybees could involve a mixture of kin selection and parental manipulation (Alexander 1974, West Eberhard 1975, Charnov 1978b, Craig 1979, 1983). One version of this scenario starts with variation in fertility among a female's daughters, caused simply by random differences in such things as genotype, larval diet, and temperature during development. Next, there was selection (through kin selection) for daughters to possess genes for helping which were facultative, only operating in subfertile females. Once these genes were widespread, the stage was set for selection on mothers to rear subfertile daughters, probably by raising larvae under suboptimal conditions, and so gain helpers. Eventually simultaneous selection on helpers to be highly efficient, and on mothers to take full advantage of this aid, produced the distinctive queen and worker castes observed today. A variant of this basic sequence is also possible, one without facultative genes for altruism and without mothers forcing altruism by rearing subfertile daughters, but simply with mothers behaviorally or perhaps pheromonally dominating their daughters when adults. Either way, the mothers might depress their daughters' reproductive potential enough to make it genetically favorable for them to stay and help rather than simply flee the maternal nest. Then, once daughters were regularly helping their mothers, they were selected to be highly efficient helpers, eventually becoming so skilled at helping that a high benefit:cost ratio for this aid "traps" them in altruism regardless of maternal domination. Finally, as in the preceding version, continued selection on daughters and mothers for effectiveness in giving and using aid led to the modern queen and worker castes of honeybees. Of these two versions for how maternal manipulation and kin selection might interact to produce sterile helpers, the first one seems more appropriate for honeybees, simply because manipulation of the nutrition of female larvae to render them helpers or nonhelpers is precisely what one observes in honeybees today.

Given that both maternal manipulation and kin selection could have helped give rise to the sterility of worker honeybees in the presence of a queen, an interesting question arises concerning the adaptive significance of the queen substance pheromone, (E)–9–oxo–2–decenoic acid (9–ODA). Exposure of workers to this substance inhibits them from rearing replacement queens and inhibits their ovaries from producing eggs (Butler 1954, Groot and Voogd 1954, Pain 1954; reviewed by Gary 1974, Michener 1974). Should we view this pheromone as a mechanism of queen domination of workers, or as a mechanism for workers to monitor the presence of their queen? In my view, both theory and observations support the latter interpretation. First, as discussed earlier, kin selection theory indicates that if the benefit:cost ratio for worker honeybee sterility exceeds about 1.8, workers are selected to forego personal reproduction. Because it is likely that this condition is fulfilled, it seems clear that queen honeybees have no need to dominate their workers continuously.

The empirical support for the view of 9–ODA as a queen-presence, rather than worker-domination, signal comes from observations on queen-worker interactions. One is the approximately 30-day lag between the experimental removal of a colony's queen and the widespread activation of workers' ovaries. This lag does not reflect a delayed drop in 9–ODA level—rearing of replacement queens commences within 24 hours of queen loss (Seeley 1979)— but rather the inhibitory effect of brood (Jay 1970, Kropáčová and Haslbachová 1970, 1971). When both the queen and brood are experimentally removed from a colony, only about 15 days must pass before over 50 percent of the workers possess well-developed ovaries. If the queen substance alone provided a means whereby queens dominate their workers, one would expect more rapid activation (in less than 30 days) of workers' ovaries following loss of the queen alone. Workers are clearly designed to begin egg-laying only if the colony lacks both a queen and brood and thus is unable to rear a replacement queen, or in other words, is hopelessly queenless. A second relevant set of observations centers on the behavior of workers toward queens. If the queen were dominating the workers with 9–ODA, one would not expect the workers to be attracted to the queen and also to help disperse the 9–ODA throughout the nest. Both behaviors are observed (Velthuis 1972, Seeley 1979, Ferguson and Free 1980). A third line of evidence comes from clues about the phylogenetic history of 9–ODA. As is discussed by Nedel (1960) and Michener (1974), mandibular glands (the source of 9–ODA) are common in both solitary and social bees, and are probably a source of sex attractant pheromones for bees in general. In honeybees, 9–ODA still functions as a sex attractant besides being a channel of queen-worker communication. In fact, the mandibular glands of queens are largest early in life, the time of queens' nuptial flights. It therefore seems fair to conclude that the primitive function of 9–ODA was signaling to males the presence of a female, and that secondarily it has acquired the function of signaling to workers the presence of their queen.

The general conclusion here is that the present-day social system of honeybee colonies is evidently not one of a despotic queen ceaselessly dominating the reproduction of thousands of worker-daughters, but rather one of workers themselves benefiting by providing for the well-being of their queen, the individual whose reproduction provides their best avenue for propagating their genes.

Labor Specialization by Workers

Division of labor among workers is central to the social organization of honeybees. The differentiation of a colony's members into labor specialists, combined with their integration through systems of communication, enables a colony to operate far more efficiently than if it were a simple aggregation of identical individuals. The principal basis for labor specialization among worker honeybees is age. During the 30-odd days in the adult life of a worker bee, each individual passes through a regular progression of tasks, starting with cell cleaning, ending with foraging, and passing through brood care and food storage in mid-life. This phenomenon of behavioral change with age is called "age polyethism." In honeybees, these behavioral changes are accompanied by regular shifts in the activity of exocrine glands. For example, as workers pass through the labor sequence of nurse, comb builder, and guard, their hypopharyngeal (brood food) glands, wax glands, and alarm pheromone glands successively rise and fall in size and secretory activity.

Past investigations of honeybee age polyethism have focused either on documenting these patterned shifts in behavior and physiology (reviewed by Free 1965, Wilson 1971, Michener 1974), or, more recently, on analyzing the endocrine mechanisms timing these shifts. The level of juvenile hormone in the haemolymph of worker bees rises steadily during adult life, evidently functioning as the central pacemaker, or at least control signal, for the age-related changes (Rutz et al. 1974, 1976, Fluri et al. 1982, Robinson 1985). An important, but little-studied, aspect of honeybee age polyethism is the adaptive significance of the honeybee's schedule of labor changes.

The essential story of honeybee age polyethism is shown in Figures 3.5 and 3.6. The first figure illustrates that although workers perform many distinct tasks during their lives, they experience only three periods of dramatic change in labor repertoire, at or around the ages of 2, 11, and 20 days. One consequence of these relatively synchronous (for bees of a given age) switches in task performance is that there exist four clearly defined age castes among workers. The first caste, a cell-cleaner caste, comprises bees of age 0–2 days. Comparison of Figures 3.5 and 3.6 reveals that the second caste, consisting of workers 2–11 days old, is a broodnest caste. All of the tasks performed by this caste occur within the central, broodnest portion of the nest, although some of their tasks (such as shaping comb and ventilating) also extend outside the broodnest. Conversely, workers in the third caste, age 11–20 days, constitute a food-storage caste, whose tasks all occur in the peripheral, food-storage region of the nest. Finally, the fourth caste contains only foragers, bees which are generally 20 or more days old and which work primarily outside the nest. In the terminology of Wilson (1976), the age castes of worker honeybees are discretized, that is, workers are organized into largely non-overlapping age groups, each of which handles a distinct set of tasks.

What is the adaptive significance of this particular labor schedule? One general property of the schedule—the progression from inside-nest to outside-nest labor—is clearly adaptive. Workers living inside the protected environs of their colony's nest experience negligible mortality, but once they begin leaving the hive to go on foraging trips, they become exposed to predation and accidents and start perishing at a very high rate (Sakagami and Fukuda 1968). By postponing the most dangerous work to late in life, a worker helps ensure a long life of aid to her mother.

The functional design of the labor schedule for inside-nest tasks is not so immediately obvious. One hypothesis is that it reflects a compromise between selection for efficiency in performing tasks and selection for efficiency in locating tasks (Seeley 1982a). The reasoning behind this hypothesis runs as follows. First, presumably a worker's task-performance efficiency would be highest if the worker performed just one task at each age. This way a worker could become highly skilled at detecting the cues relevant to each task and, through concentrated practice, could perfect her responses to these cues. That such task-specific learning occurs is clear. Foragers, for example, must master the use of certain orientation cues, such as the direction of the sun's movement across the sky (Lindauer 1959), and the relationship between sun azimuth and landmarks about the nest (Dyer and Gould 1981). Comparable learning for inside-nest tasks probably also occurs, but has never been carefully analyzed. Preliminary evidence of this learning consists of reports of a few

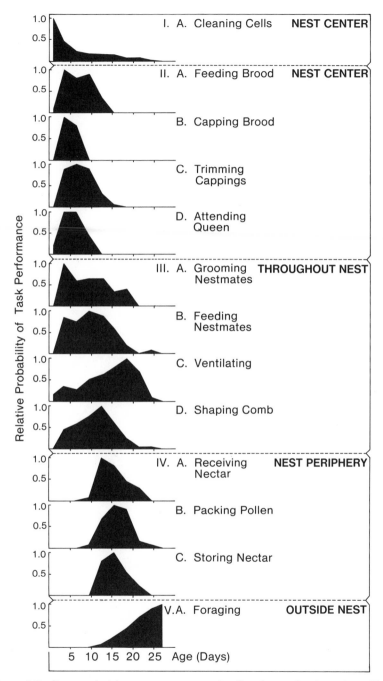

Figure 3.5 Changes in labor patterns across the life of a worker honeybee. (From Seeley 1982a.)

I. A. CLEANING CELLS

II. A. FEEDING BROOD

B. CAPPING BROOD

C. TRIMMING CAPPINGS

D. ATTENDING QUEEN

III. A. GROOMING NESTMATES

B. FEEDING NESTMATES

C. VENTILATING

D. SHAPING COMB

IV. A. RECEIVING NECTAR

B. PACKING POLLEN

C. STORING NECTAR

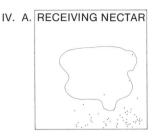
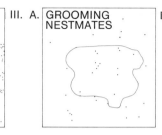
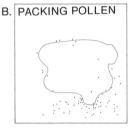
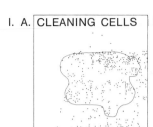
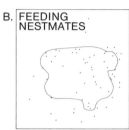
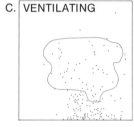
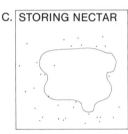
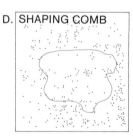

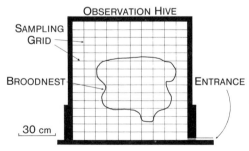

OBSERVATION HIVE

SAMPLING GRID

BROODNEST

ENTRANCE

30 cm

workers repeatedly performing rare tasks, such as guarding and dead bee removal (Visscher 1983), and of workers improving their proficiency in intricate tasks, such as following recruitment dances (Lindauer 1952). However, if workers were to specialize on one task at each age, they would probably experience low task-location efficiency because workers would have to search relatively widely for additional work sites of their single current task. By this reasoning, one expects that workers will not be extreme labor specialists, but instead will be semi-specialists, performing a set of tasks at each age. Such versatility is well documented by the reports of Lindauer (1952), Sakagami (1953), and others who have followed individually identifiable workers for extended periods and observed them perform several distinct tasks in succession. But what determines which tasks will be performed together at each age? Again, task-location efficiency may be important. If the tasks performed concurrently also co-occur spatially in the nest, then the mean free path between tasks should be minimized, and this should help maximize efficiency in locating tasks.

To test this spatial-efficiency hypothesis one can check whether the task-set for each age of pre-foraging workers maps onto a specific nest region, or in contradiction to the hypothesis, onto spatially segregated sites about the nest. The labor patterns shown in Figures 3.5 and 3.6 support the spatial-efficiency hypothesis. Caste 1 workers perform a single task, cell cleaning, so there is no question of different tasks in different regions. Caste 2 (broodnest) and caste 3 (food storage) workers can perform all their tasks within the boundaries of the broodnest and food storage areas, respectively. A second, more stringent, test of the spatial-efficiency hypothesis involves continuously following individual workers, observing whether at each age they do in fact specialize on the tasks occurring within a particular nest region. When Nowogrodzki (1983) recently conducted such observations, he found, in contradiction to the spatial-efficiency hypothesis, that workers spend time in various regions of their nest on all but the first few days of adult life, when they reside almost exclusively in the broodnest. However, Nowogrodzki worked with bees living in a small observation hive in which the broodnest and food storage areas were much smaller and perhaps less well segregated than in a natural honeybee nest. Further studies at the level of individual honeybees are needed to understand the importance of nest architecture and spatial efficiency in shaping the honeybee's age polyethism schedule.

Division of labor among workers also occurs on a much finer scale than the four, broad age classes discussed so far. This is illustrated most vividly within the forager caste, where individuals specialize on different species of

Figure 3.6 Maps depicting the work sites for the 12 most common tasks performed inside the nest. Lower right: schematic diagram of the observation hive used in mapping the work sites of the various tasks. (From Seeley 1982a.)

forage plants, on different patches of flowers of a particular species, and even on different foods (pollen or nectar) from a particular patch of flowers (reviewed by Ribbands 1953, Free 1970). For example, although a colony as a whole might be collecting 10 or more pollen types at any one time, each pollen forager tends to remain faithful to one species. In one study (Free 1963), 70 to 90 percent of pollen foragers stayed with one species for a day, and 40 to 60 percent remained constant to a species for a week. Likewise, Singh (1950) and Weaver (1957) plotted the courses of labeled foragers on golden rod (*Solidago virgaurea*), hairy vetch (*Vicia villosa*), buckwheat (*Fagopyrum esculentum*), and other species and found that individuals frequently foraged over the same 50 m² or smaller area for several days. Parallel phenomena for members of the food-storage (nest periphery) caste include bees which specialize for several days as guards, water collectors (Lindauer 1952, Robinson et al. 1984), or undertakers (Sakagami 1953, Visscher 1983). Such specialization within age castes probably reflects learning of the special skills required for certain tasks—how to enter and manipulate flowers of a certain design, the layout of especially rich flowers in a patch, the peculiar odor of dead bees, the location of a nearby water source—and then focusing on these tasks for maximal labor efficiency.

Colony Life Cycle

The life of a honeybee colony can be regarded as beginning when a strong, established colony starts rearing a batch of queens in preparation for swarming (Fig. 3.7). The first event is the construction of queen cups along the lower margins of the colony's broodnest combs (Allen 1965a). These structures, tiny inverted bowls made of beeswax, form the bases of the large, ellipsoidal cells in which queens are reared. Next the queen lays eggs in twenty or more of the queen cups and workers feed the hatching larvae the royal jelly which ensures their development into queens. The formation of these new queens is remarkably rapid, requiring only 16 days from the time the egg is laid to the moment when an adult queen emerges from her cell. As the daughter queens develop, changes unfold simultaneously in the physiology of the colony's queen. With each passing day, she is fed less and less by the workers. Her egg production declines and her abdomen, no longer swollen with fully-formed eggs, shrinks dramatically. Furthermore, the workers begin to shake their queen, pressing on her with their front legs or head and letting loose a burst of five or six shaking movements (Milum 1955). Such shakings, which can eventually reach a frequency of 40 to 80 bouts per hour, appear to force the queen to keep walking about the nest. This exercise, together with reduced feeding, results in a 25 percent reduction in the queen's body weight (Allen

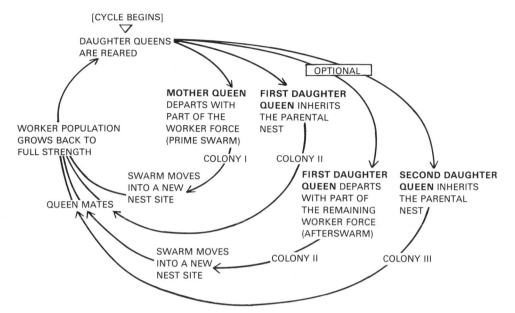

Figure 3.7 Principal events in the life cycle of honeybee colonies. (Modified from Wilson 1971.)

1960, Fell 1977). Shortly after the first queen cell is capped, the mother queen flies off in a swarm of some ten to fifteen thousand workers, leaving about the same number of workers behind in the parental nest. After flying a short distance, the swarm condenses into a beard-like cluster on a tree branch. From here the swarm's scout bees explore for nest cavities, select one which is suitable, and finally signal the swarm to break cluster and fly to the new home site.

For about eight days following the departure of their mother queen, the workers in the parental nest are queenless, but this situation ends with the emergence of the first virgin queen. If the first (so-called "prime") swarm's departure has greatly weakened the parental colony, then the remaining workers allow the virgin queen that emerges first to search through the nest for her rival sister queens and to kill them while still in their cells. Frequently, however, by the time the first virgin queen has appeared, sufficient worker brood has also emerged to restore the parent colony's strength. In this situation, the workers guard the remaining queen cells against destruction by the first virgin queen, start shaking this queen in preparation for flight, and eventually push her out of the nest in an "afterswarm." This process is repeated with each emerging queen until the colony is weakened to the point where it cannot support further fissioning. If more than one virgin queen now remains in the parental nest, the workers simply allow these queens to fight each other until just one queen remains. Because the occurrence of after-swarms depends on the colony's initial strength in workers and brood, the number of afterswarms per bout of colony reproduction varies tremendously among colonies (Huber 1792, Allen 1956, Winston 1980).

The reproductive process is completed when the surviving virgin queens fly out of their nests to mate with males from the surrounding colonies, the colonies in new nest sites construct their nests, and all of the colonies strive to quickly rebuild their populations and store enough honey for winter survival. Under favorable conditions, each colony will live to repeat the entire process the following summer. Death eventually comes to colonies through starvation, predation or disease, or failure to replace a senescent queen.

4

The Annual Cycle
of Colonies

Introduction

One key to understanding the ecology of honeybees living in cold climates is the unique annual cycle of their colonies. In winter, when colonies of the other social bees (bumblebees and social sweat bees) have dwindled away, leaving only a residue of fertilized females deep in hibernation, honeybee colonies continue to thrive as true social groups, each one maintaining a labor force of some fifteen thousand workers. Moreover, rather than entering chilly dormancy, as is the rule for insects in cold climates, honeybee colonies resist the cold, regulating the temperature of the colony perimeter above about 10°C even in ambient temperatures of −30°C or colder. To achieve such temperature control, honeybees nest inside protective cavities, press tightly together to form a well-insulated cluster, and pool the metabolic heat generated by microvibrations of their powerful flight muscles. The fuel for this winter-long heating process is the twenty or so kilograms of honey stockpiled by the colony over the previous summer.

The honeybee's annual cycle is unique in other ways besides the overwintering process. Shortly after the winter solstice, when the days grow longer but snow still blankets the countryside, each honeybee colony raises the core temperature of its winter cluster to about 34°C and starts to rear brood. At first, only a hundred or so young bees are produced, but by early spring, when the first flowers blossom, over one thousand cells hold developing bees and the pace of colony growth quickens daily. Come late spring, when bumblebee and sweat bee queens are just rearing their first daughter-workers to adulthood, honeybee colonies have already expanded to full size, thirty thousand or so individuals, and have begun to reproduce. Reproduction involves not only the standard process of rearing males, which simply fly from the nest and mate, but also an intricate process of colony fission in which a labor force of several thousand workers accompanies each departing queen.

One aim of this chapter is to provide a macroscopic view of honeybee life by describing the broad pattern of events which unfolds in honeybee colonies each year. Such a view reveals several fundamental themes of honeybee ecology, themes which help unify the subsequent chapters. A second goal of this chapter is to try to explain the evolutionary origins of the honeybee's unique annual cycle. As we shall see, the honeybee's distinctive ecology reflects a curious blend of historical factors and ecological forces, as novel adaptations to a harsh physical environment were superimposed on the physiological and social characteristics which this bee inherited from its tropical ancestors.

Annual Cycle of Energy Intake and Expenditure

Winter survival through thermoregulation is energetically expensive. In the middle of winter, a honeybee colony weighs approximately two kilograms (Avitabile 1978) and consumes energy at a rate of 20 to 40 watts (Southwick 1982), roughly the same rate as is burned by a small incandescent lamp. Naturally, a colony's long-term survival requires that this rapid wintertime energy expenditure be balanced by intensive energy storage during the warm summer months, when the rich nectar of flowers can be harvested to rebuild a colony's honey stores. A high flow of energy between colony and environment is therefore a fundamental feature of honeybee ecology, and one which provides ecologists with a valuable window on the lives of bees. By monitoring this energy flow throughout a year, one gains not only a broad, abstract view of the honeybee's annual cycle, but also a quantitative picture of one of the honeybee's greatest problems: the need to amass within a short summer season an ample supply of winter heating fuel.

Precise measurements of net energy flux through a honeybee colony have so far always involved colonies sealed inside an air-tight chamber (Free and Simpson 1963, Heinrich 1981a, Kronenberg and Heller 1982, Southwick 1982). This allows precise measurement of colony metabolic rate through analysis of carbon dioxide production or oxygen consumption, but of course prevents bees from flying to and from the colony. Thus these measurements provide a good indication of energy flow out of colonies at night, or when ambient temperatures are low, but they cannot provide a complete picture of colony energetics, which must include the situation of a colony accumulating energy through foraging. A less exact, but far more versatile, method of monitoring the net energy flow through a honeybee colony is to record changes in total colony weight (the weight of the bees, the nest, and the food stored within the nest) (McLellan 1977). Weight is gained when forage is brought into a colony, but is lost when stored food is metabolized, when a colony divides for reproduction, or when colony members die.

Records of weight changes in honeybee colonies across a summer or throughout a year have been published for many regions of the world, including the United States (Milum 1956, Oertel 1958, Thompson 1960, Visscher and Seeley 1982), Canada (Mitchener 1955), West Germany (Koch

1967), and England (Crane 1975). Almost without exception, however, these records were collected for apicultural, not ecological, purposes and so reflect the energetic patterns of beekeepers' colonies managed for honey production and living in agricultural settings. Therefore, the discussion which follows is based primarily on the findings of just one study (Seeley and Visscher 1985), which used undisturbed colonies inhabiting hives whose volume (84 liters) resembles those of the natural nests of honeybees. The least natural aspect of the ecology of these colonies was their habitat: a botanical garden on the edge of New Haven, Connecticut. Because forage is probably richer in this setting than in a more natural environment such as an undisturbed forest, the results of this study almost certainly underestimate the challenge of food collection faced by honeybees living in nature.

The weight records shown in Figure 4.1 illustrate that for honeybee colonies winter is a time of dramatic weight loss. On average, the colonies studied lost 23.6 ± 2.8 kg (mean ± one standard deviation). Except for approximately one kilogram, attributable to the removal of dead workers from the nest (Avitabile 1978), all of this weight loss represents consumption of stored food—honey and pollen. Probably the actual weight of food consumed is a good deal higher than the above numbers suggest, since in the late winter and early spring a colony converts much of its food reserve into workers (see below). In fact, much of the 20+ kg cost of overwintering evidently reflects the high cost of winter brood rearing. One indication of this is the steady rise in weekly weight loss as the winter progresses and brood rearing intensifies. For example, colonies lost weight twice as rapidly in March (−0.84 ± 0.25 kg/week) as in December (−0.42 ± 0.12 kg/week). A second indication comes from the experiments of Farrar (1934, 1936), who compared the winter weight losses of colonies with and without stored pollen (hence with and without winter brood rearing). The weights of colonies with pollen fell 22.7 kg, on average, between October and May, whereas those of colonies lacking pollen dropped only 11.8 kg over the same period. Presumably these differences in weight loss pattern stem primarily from the change in the colony's thermoregulation strategy when it commences brood rearing (reviewed by Seeley and Heinrich 1981; also see Chapter 8). Whereas a colony without brood simply maintains its surface temperature at about 10°C, thus just a few degrees above the honeybee's chill-coma threshold, a colony with brood must maintain its core at 34°C, the honeybee's broodnest temperature.

A second important generalization illustrated by Figure 4.1 is that the annual period of net energy intake is brief. The colonies studied by Seeley and Visscher (1985) only gained weight for 14.0 ± 1.7 weeks each year. What is even more striking, in these colonies 86 percent of the annual weight gain occurred between April 16 and June 30—a period of just 75 days.

The theme which emerges from the analysis of colony energetics across a

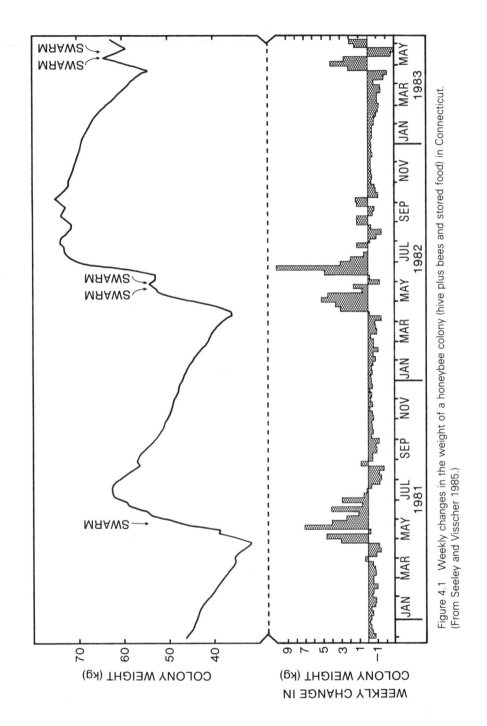

Figure 4.1 Weekly changes in the weight of a honeybee colony (hive plus bees and stored food) in Connecticut. (From Seeley and Visscher 1985.)

year is one of a constantly threatening energy crisis. A colony consumes 20 or more kilograms of honey each winter, yet has little time each summer in which to assemble its food reserves. As we shall see in the next section, this energy problem is especially severe for newly founded colonies, which, unlike colonies which are already established, have no opportunity to fall back on reserves from previous years, and moreover face the high costs of nest construction. Certainly not all honeybee colonies experience such severe energy problems. Colonies at lower latitudes or in habitats richer in forage than those discussed above may well find other problems, such as predators or nest site shortages, far more challenging. Still, for colonies living in nature throughout the honeybee's northern distribution, the primary challege to survival probably is the balancing of the annual energy budget.

Annual Cycles of Colony Growth and Reproduction

The proper timing of colony growth and reproduction is essential to honeybee survival in cold climates. As we have seen, colonies survive winter through thermoregulation, and to do so must amass tens of kilograms of honey during the summer. By properly adjusting the timing of growth and reproduction in the colony's annual cycle, honeybees help ensure that the foraging force can collect the essential winter stores before the summer ends.

The patterns of colony growth and reproduction across a year are both easily quantified. Colony growth patterns have been described either by periodically measuring the number of brood-filled cells in a colony's nest (Nolan 1925, Allen and Jeffree 1956, Jeffree 1956, Avitabile 1978, Winston 1981) or by repeatedly censusing a colony's population of adult bees (Bodenheimer 1937, Jeffree 1955). Describing the annual cycle of colony reproduction is slightly more difficult because it involves distinct processes for females (queens) and for males (drones). The seasonal pattern of reproduction along the male line is most easily documented by counting the cells containing developing drones. The annual cycle of reproduction through females could likewise be determined by monitoring the appearance of queen cells, the strikingly large cells which cradle developing queens. However, because colonies frequently make false starts in queen rearing, building queen cells but then tearing them down before the developing queens mature (Simpson 1957b, Gary and Morse 1962), a superior technique is to record the appearance of swarms, the true units of female reproduction in honeybees.

Figure 4.2 shows the growth pattern of honeybee colonies in Connecticut, based on bimonthly censuses of brood throughout several years. Following several months without brood in autumn and early winter, colonies begin rearing bees in January or February, evidently in response to increases in

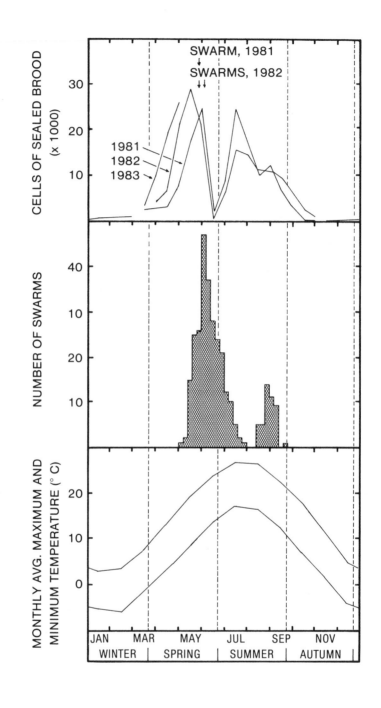

daylength (Kefuss 1978). At first, fewer than 1000 cells containing brood are found in a nest, but in late March or April this number soars, climbing to a peak of some 30,000 or more developing bees per colony in May or June. Shortly thereafter there appears a gap in brood rearing when, because of the turnover in queens associated with swarming, the colony lacks an egg-laying queen for 10 to 20 days. Once the new queen eliminates her rivals and completes her mating, full-scale brood rearing resumes, to decline only gradually throughout the remainder of the summer and finally cease altogether in October. This annual pattern holds for honeybee colonies in temperate climates (Nolan 1925, Jeffree 1955, 1956, Allen and Jeffree 1956), although the precise timing of the rapid springtime expansion varies markedly with latitude. For example, the colonies studied by Nolan (1925), located in Somerset, Maryland, entered the intensive growth phase three weeks before the time observed for colonies in New Haven, Connecticut, 250 kilometers to the north. Such geographical differences in annual cycle of brood rearing are partly under genetic control, reflecting adaptation to the local climate and flora. Colony transplant experiments conducted in France have revealed that colonies exchanged between the regions of Paris and Bordeaux, where the principal peaks in brood rearing come in early and late summer, respectively, maintain their distinctive annual cycles of brood rearing in their new habitats (Louveaux 1973).

Reproduction by honeybees commences in late spring, shortly after the abrupt rise in brood rearing, and is largely completed by the start of summer. For example, in southern England, drones were found to constitute 9 percent of a colony's brood in May and June, but only 1 percent in July and August (Free and Williams 1975), in correspondence with the general spring schedule of drone production in north temperate regions (Allen 1958, 1965a, 1965b, Weiss 1962, Page 1981). Swarm production coincides with the appearance of drones. Figure 4.2 shows, for example, that of 301 swarms collected in central New York State between 1971 and 1981, 84 percent were found between May 15 and July 15. Numerous reports from throughout North America and Europe confirm this pattern of late spring–early summer swarming (Mitchener 1948, Jeffree 1951, Murray and Jeffree 1955, Simpson 1959, Fell et al. 1977, Caron 1980, Page 1982), although, as with brood rearing, there exists geographical variation in the peak time of swarming.

Figure 4.2 Annual cycles of honeybee brood rearing and of air temperature in Connecticut, and of honeybee swarming in New York State. Top: counts of the number of cells of brood per colony throughout the year. Middle: seasonal occurrence of 301 swarms collected over the ten-year period of 1971 to 1981. Bottom: annual pattern of air temperature for New Haven, Connecticut. (From Seeley and Visscher 1985.)

Few features of honeybee biology are more impressive than the ability to initiate colony growth in the middle of winter, long before any flowers provide fresh forage, and thereby build up a colony's population to swarming strength by spring or early summer. The ecological significance of this remarkable skill is that it provides newly founded colonies with sufficient time to construct nests, raise new workers, and store winter provisions before winter's cold calls an end to foraging. One measure of the energetic hurdle faced by new colonies is their low probability of survival over the first winter. In New York State, only 24 percent of the new colonies each summer remain alive the following spring, whereas fully 78 percent of already established colonies survive from year to year (Seeley 1978).

A recent set of experimental studies has underscored the critical significance of early colony growth and reproduction to honeybee survival in cold climates (Seeley and Visscher 1985). In one experiment, colonies whose onset of brood rearing was experimentally postponed from mid-winter to April 15 (when fresh forage became available) were compared to colonies allowed to rear brood on the normal schedule. Whereas the control colonies contained $10,800 \pm 2800$ bees by May 1, the experimental colonies contained only 2600 ± 1100 bees. Furthermore, the colonies that reared brood in the winter swarmed considerably earlier than those of the other group (mid May to late May vs. late June to early July). A second experiment examined the importance of early swarming. Here a comparison in the probability of winter survival was drawn between colonies started on May 20 and June 30, thus 20 days before and 20 days after the median swarm date in New York State. In three out of the four years studied, forage was either extremely rich or unusually poor (as measured by weighing the hives in the autumn), and either all colonies died (in the poor year) or nearly all survived (in the two rich years). However, in the one year with only moderate forage, nearly all early swarms survived whereas all late swarms perished.

Evolution of the Annual Cycle

Why does the honeybee's annual cycle differ so strongly from those of the other social bees inhabiting cold climates, such as the bumblebees? As we have seen, honeybees create year-round colonies, maintain a warm microclimate inside their nest throughout winter, time the growth and reproduction of their colonies to peak early in the summer, and reproduce by swarming. In bumblebees, by contrast, colonies are founded in spring by lone queens, grow dramatically in the summer, and reproduce in late summer by rearing solitary reproductives, the females of which mate and overwinter singly (Michener 1974, Alford 1975). It has been argued above that the honeybee's yearly cycle is a tightly interwoven set of traits through which honeybees survive the harsh winters of temperate regions. At first glance, then, the honeybee's unusual way of life might seem to reflect an advanced, superior technique of insect overwintering, one based on active control of the physical environment rather than on circumventing the threat of freezing temperatures through passive dormancy. However, this is clearly not the case. The bumblebee's winter survival system, although far simpler than the honeybee's, allows it to survive harsher winters. Worldwide, the northernmost limit of bumblebees extends thousands of miles beyond that of the honeybee. Bumblebees even thrive in tundra north of the Arctic Circle (Richards 1973, Michener 1974). The colony cycle differences between honeybees and bumblebees are evidently rooted in the distinct geographical settings in which their societies arose: the warm tropics for honeybees and the cool temperate regions for bumblebees (Michener 1979).

A comparison of colony cycles between the two major groups of social bees in the tropics—the honeybees (Apini) and the stingless bees (Meliponini)—reveals two major traits held in common: colony lifespans of several years and reproduction by colony fission. Undoubtedly the perennial nature of these bees' colonies reflects evolution in mild tropical climates; there is no need to have a hibernating, solitary phase in the colony cycle. Colony multiplication through swarming is probably an outcome of intense predation pressure in the tropics. Ant predation on social insects, for example, intensifies strongly as one travels from temperate to tropical latitudes. When Jeanne (1979) timed the disappearance rate of social wasp larvae placed in vials accessible only to ants, he found that in Costa Rica (10°N) less than one-seventh of the larvae remained alive after 48 hours, but in New Hampshire (43°N) more than one-half survived. In an environment where predation is severe, there would obviously be strong selection for colony-founding queens to be accompanied by workers. The presence of workers not only ensures continuous guarding of the new nest, but also protects the queen by freeing her from making hazardous foraging trips. Without the need for behavioral totipotency that is required of insects that initiate colonies singly, honeybee queens evidently regressed in evolution toward the role of simple egg-laying machines. The workers, likewise, evolved their own distinct set of specialized morphological, physiological, and behavioral traits. Ultimately there arose the absolute interdependence between queen and worker castes found in honeybees today.

Thus when the honeybees expanded out of the tropics into cold temperate regions, their adaptation to a harsh physical climate was severely constrained by their advanced social organization. Instead of overhauling this social system in order to overwinter in bumblebee fashion, as solitary, totipotent queens, or imposing the physiological modifications which would enable entire col-

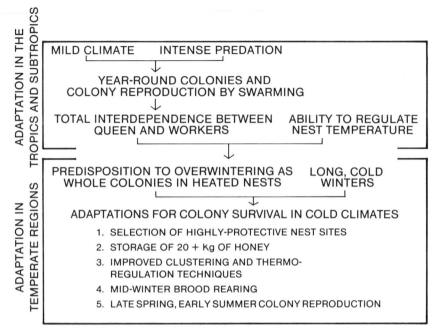

Figure 4.3   Inferred history of the evolution of the honeybee's distinctive annual cycle.

onies to overwinter in dormancy, honeybee evolution pursued what was perhaps the simplest route to winter survival: refinement of such pre-existing
techniques as colonial thermoregulation, food storage, nest site selection, and
seasonal control of colony growth and reproduction. In this way honeybees
came to overwinter as whole colonies without actually hibernating. Figure
4.3 summarizes this evolutionary scenario. Bumblebees, in contrast, probably
evolved their social life in a cool climate after evolving the habit of winter
survival through dormancy. Thus bumblebees appear to have built their social
organization around their mode of overwintering, rather than vice versa, as
occurred with honeybees.

5 Reproduction

Introduction

Several of the knottier puzzles in the biology of honeybees emerge upon considering the evolution of their patterns of reproduction. Here we seek to understand, for example, how natural selection favors the immensely skewed (in favor of males) sex ratio, the multiple-year lifespan of queens, and the size and number of swarms produced annually by a colony. The difficulty in understanding such patterns stems primarily from the twin facts that honeybees live in and reproduce through colonies comprising thousands of individuals, yet natural selection operates mainly at the level of individual colony members. Thus, on the one hand, we expect the members of a colony to cooperate with one another to a high degree, since overall colony success is a prerequisite to any individual's reproductive success. On the other hand, because each colony member possesses a unique genotype, we should also expect considerable conflict among individuals over the use of a colony's resources in reproduction. Thus the particular patterns observed in honeybee reproduction are likely to reflect a complex mixture of cooperation, conflict, and compromise. The approach adopted here for unraveling the tangled forces shaping particular reproductive patterns is to define for colony members in different roles (worker, mother queen, daughter queen) the genetic interest in the outcome of a given reproductive process, and then to analyze the observed patterns to identify which individuals exert the greatest influence on each aspect of reproduction.

Investment Ratio Between Queens and Drones

One of the most intriguing features of honeybee reproductive biology is their astonishingly skewed sex ratio. Each year, a typical colony rears to adulthood between 5000 and 20,000 males (Allen 1965a, 1965b, Weiss 1962, Page 1981), but only about 10 queens, of which only 1 to 4 will mate and head a colony (Huber 1792, Allen 1956, Winston 1980). Current theory, however, predicts that in the simplest case (large population, homogeneous individuals, random mating, etc.), natural selection will favor parents who allocate equal resources to male and female reproductives (Fisher 1930, Charnov 1978a, 1982).

As a first step toward resolving this apparent incongruity between theory and observation, let us consider what constitutes a colony's total investment in each sex. For males, this is straightforward. It is simply the cost of rearing drones plus the cost of supporting them throughout adulthood. For females, the situation is more complex. It seems correct to count as investment in female reproductives everything, except the drones and the resources needed to support them, which is left behind in the parental nest after the prime swarm (containing the mother queen) departs (Fig. 5.1). This includes the daughter queens, the workers which stay behind, the wax combs, and most

Figure 5.1   A swarm of honeybees, consisting of one queen and approximately 12,000 workers. These bees have recently left their old nest and have settled on these branches to rest quietly until the swarm's scouts have located a new home site.

of any honey and pollen which they hold (Hamilton 1975). Each of these items represents a resource which helps the daughter queens survive and reproduce, and which is given to the daughter queens at a cost to the mother colony (the mother queen and her accompanying workers). For example, if the nest were not given to one of the daughter queens, but instead was kept by the mother colony, then presumably the mother colony would need fewer workers and so could invest more workers in the daughter queens. The key idea here is that a colony possesses a finite stock of resources (workers, stored food, nest materials) which it can partition between reproduction and self-survival. If during reproduction a given resource is not employed in fostering mother-colony survival, it is expended in colony reproduction, and vice versa.

Given the preceding definitions of investment in male and female reproductives, we see that what determines a colony's investment ratio between male and female functions is its intensity of drone production and its pattern of self-division when the prime swarm leaves. Now let us consider the preferences of different colony members regarding the ratio of sex allocation. Probably only the preferences of the mother queen and her workers are important here, since it seems unlikely that either drones or daughter queens can influence their colony's investment patterns. Drones neither contribute to further drone production, nor can they influence colony fissioning. Likewise, the new queens only emerge as adults after all the important decisions shaping the sex investment ratio have been made.

It seems likely that the mother queen of a colony has been shaped by natural selection to favor an equal (1:1) ratio of overall investment in male and female reproductives. The queen is equally related ($r = 0.50$) to her sons and daughters, and in honeybees there does not appear to be any spatial structuring of populations which would select for a biased ratio of investment. In particular, there is probably little or no local mate competition (Hamilton 1967, 1979) since both males and females fly a couple kilometers or more (see below) before mating, thereby ensuring genetic mixing among colonies and thus minimizing competition among siblings for mates. It has been suggested that the reproductive investment patterns of social insects, such as honeybees and army ants, in which female reproduction involves colony fissioning, represent the result of local resource competition (Stubblefield 1980, Craig 1980) in which selection should favor male-biased investment ratios. I believe that for honeybees, at least, this is not the case. The key question concerning local resource competition is whether there is a decreasing return in reproductive gain per unit investment in a sex as investment in that sex increases (Charnov 1982). If one defines investment in females as I have done above, to include the total resources received by all daughter queens including those in afterswarms, one sees that there is probably little or no diminution in return to female investment. Further investment in females probably provides an approximately linear increase in reproductive gain, either by improving the survival probability of a given number of colonies headed by daughter queens (by making the swarms bigger), or by increasing the number of such colonies.

Workers, unlike queens, are not equally related to their colony's male and female reproductives and so apparently prefer a ratio of sex allocation somewhat different from 1:1 (Trivers and Hare 1976, Charnov 1978a). If a worker's mother queen is singly inseminated and lays all the male eggs in the colony, then the worker will be related by three-fourths ($r = 0.75$) to the new queens, but only by one-fourth ($r = 0.25$) to the males. In this situation and given the absence of spatial structuring of populations, as described above, workers will be selected to have a 1:3 (male:female) investment ratio, since with this investment pattern the greater genetic return on a unit of female reproduction, relative to male reproduction, is precisely canceled by its greater cost. In fact, although honeybee queens do lay all of the male eggs in a colony, they are inseminated by more than one male (see Chapter 3). Charnov (1978a) has shown that in this situation (assuming outbreeding in a uniform environment), the workers' preferred fraction of investment in females is $(n + 2)/(2n + 2)$ and in males is $n/(2n + 2)$, where n represents the number of males inseminating the queen. These fractions also assume that all n males have an equal probability of being the father of any particular queen produced by the colony (i.e., all males contribute an equal number of sperm, the sperm are chosen at random for fertilization, and the workers do not bias the patriline representations among the new queens). The equilibrium investment ratios predicted by these formulae for various values of n are shown in Table 5.1.

Since it is the workers that rear a colony's brood and define a colony's fissioning pattern when the prime swarm leaves, it seems the workers can

Table 5.1

Equilibrium ratios of sex investment for honeybee workers as a function of the number of males inseminating the queen.

Number of males	Investment ratio (M:F)
1	1 : 3.00
2	1 : 2.00
4	1 : 1.50
6	1 : 1.33
8	1 : 1.25
10	1 : 1.20
15	1 : 1.13
20	1 : 1.10

influence the ratio of sex allocation more than the queen can. However, as has been suggested by Craig (1980), the overall results may be a compromise between queen and worker preferences. The workers should want less investment in males than the queen does, and the workers could destroy male eggs or larvae to bias the investment pattern. The queen, in turn, could retaliate by laying more male eggs, which would set the stage for further male destruction by the workers. This process of retaliation and counter-retaliation would eventually have to stop since at some point the opposing actions of the queen and workers would result in a lowering of their respective fitnesses more than achieving the "correct" sex ratio would raise them.

What is the ratio of investment between males and females in honeybee colonies? Unfortunately, this has never been measured with precision. The only serious attempt to estimate this value is that of Macevicz (1979), who collated observations from many sources to calculate estimated sex investment ratios (male:female) of 1:15.8 and 1:2.6. (The first estimate includes the parent colony's nest in the female investment; the second estimate does not.) These numbers must be viewed as the crudest of estimates, but they do at least suggest there is a female bias in the investment ratio, unlike the sex ratio itself which is vastly male-biased. Future attempts at estimating the honeybee's sexual investment ratio should be based on measurements taken only from colonies living as undisturbed as possible. It should also be kept in mind that the theoretically predicted investment ratios are population properties, not colony properties. This is critical because it seems likely that honeybee colonies show facultative sex ratios such that when they are weak they produce only males (low cost per reproductive) and that when they are strong they "overinvest" in females (high cost per reproductive). (See Trivers and Willard 1973 and Charnov 1982 for general discussions of facultative sex ratios.) It is well known that honeybee colonies will produce drones without swarming (Allen 1965b). Thus a proper estimate of sex allocation in honeybees must monitor the reproductive activities of a population of colonies, weak and strong alike.

Despite the lack of data which would allow an accurate estimation of the sexual investment ratio, several sets of observations exist which suggest that colonies closely regulate their investment in males and females. For one, the percentage of a colony's adult bees which depart in the prime swarm is fairly uniform: Martin (1963) reports 67 ± 13 percent (mean $\pm$ one standard deviation) departing bees, based on 9 colonies. Second, colonies appear to take several steps to regulate their drone production. When allowed to construct nests on their own, colonies devote on average 13 to 17 percent of the total comb area to drone comb, which has larger cells for rearing the heavy-bodied drones (Weiss 1962, Seeley and Morse 1976) (Fig. 5.2). Experiments have shown that this regulation occurs through negative feedback on further

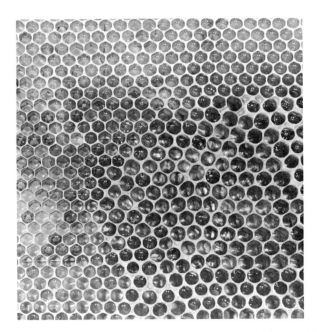

Figure 5.2 The large diameter cells in which drones are reared (lower right) contrast markedly with the smaller cells used for rearing workers (top and left).

drone comb construction by the presence of existing drone comb, though the precise mechanism of this feedback is unknown (Free 1967a, Free and Williams 1975). A second control on drone production evidently operates during the nursing of drone brood. Colonies provided with a surplus of drone comb nevertheless show an upper limit in the amount of drone brood reared (Allen 1963, 1965b). An oversupply or undersupply of developing drones somehow provides negative or positive feedback on further production (Free and Williams 1975).

Paternity of Virgin Queens

One point of possible conflict among the workers in a colony concerns the question of which female larvae will produce the next generation of queens. The basis for this conflict is the multiple paternity of a colony's females; workers are much more closely related to queens which are their full sisters ($r = 0.75$) than to those which are their half sisters ($r = 0.25$). All else being equal, workers might be expected to manipulate the investment in developing queens so that their full sisters become the future queens. For this to happen there are at least two requirements: (1) workers be able to distinguish between full and half sisters, and (2) workers of some patrilineal groups dominate the workers of other groups throughout the process of queen rearing and queen selection. If honeybee workers can distinguish full sisters from half sisters as brood or adults, it almost certainly occurs through phenotype matching or recognition alleles (probably the former, Hölldobler and Michener 1980), since it is likely that alternative mechanisms of kin recognition (by spatial distribution or association, Holmes and Sherman 1982, 1983) are precluded by extensive intermingling of the patrilineal groups. One theoretical analysis (Getz 1981) indicates that for organisms with haplodiploid sex determination, such as honeybees, phenotype matching may be several times more efficient than for diplodiploid organisms, but nevertheless evidently requires considerable genetic information. For example, if female honeybees should possess certain alleles, each of which is expressed as a distinct label (most likely an odor), and workers can perceive the number of labels that it shares with another female, then at least about 30 alleles, spread among 3 to 6 independent loci, are needed to get the expected error for full-sister versus half-sister discrimination below 10 percent.

What evidence is there that workers really can distinguish full-sister and half-sister queens? Boch and Morse (1974, 1979) have demonstrated that workers can distinguish their mother queen from an unrelated queen, even if both queens have shared the same nest and so presumably carry the same colony odor. This suggests that workers can discriminate among individual queens based on distinctive, genetically controlled, odors, but one wonders if these odors would be sufficiently different to enable discrimination if the queens were as closely related as sisters. The studies of Breed (1981) showed a positive correlation between the genetic similarity of two queens and the probability of one queen being swapped for the other without the workers of the first queen killing the substitute queen. These results suggest that workers can discriminate between individual queens related to each other as half sisters (colony-odor effects were precluded by rearing all queens used in each test in the same nest). Somewhat superior evidence for full- and half-sister discrimination exists for the situation of the sisters being workers instead of queens. Getz et al. (1982) report that when two colonies swarmed, and each colony contained just two patrilines, in each case the workers showed a tendency to segregate along kin lines. For example, from a colony which consisted of 54 percent workers from one particular patriline (recognized as "cordovan" bees, bees which carry a mutant gene for light-brown cuticle), the swarm and the remnant fraction in the parent nest contained, respectively, 58 and 43 percent bees of that patriline. However, these relatively small differences could simply reflect a difference between cordovan and the non-cordovan bees in their tendency to leave the parent nest during swarming.

Somewhat more convincing is the observation that when individual workers are introduced to a small group of workers, and all the workers in the group come from adjacent cells in a brood comb, an introduced worker is less likely to be bitten if she is a full sister than if she is a half sister of the receiving workers (Getz and Smith 1983) (Fig. 5.3). However, it is not clear that this discrimination actually occurs in nature. In the tests of Getz and Smith (1983), workers were exposed only to full sisters for five days before being presented with a strange worker, either a full sister or half sister. If the workers' recognition of kinship differences reflected habituation to the odor of their groupmates, rather than their own odor, then such discrimination probably would not occur under normal conditions, where kin groups intermingle and workers will habituate to the odors of both full and half sisters. A true test for full-sister / half-sister discrimination will involve observing the behavior of workers toward full and half sibs in a normal colony living undisturbed in an observation hive. Sherman's (1980, 1981) studies of kin recognition in Belding's ground squirrels (*Spermophilus beldingi*) exemplify the proper approach. Individuals are initially labeled for individual identification, next the patterns of behavioral interaction among individuals are recorded, and finally the labeled individuals are analyzed to determine kinship patterns and a check is made for correlations between kinship and behavior. This approach relies upon a means of a posteriori determination of kinship patterns which has the

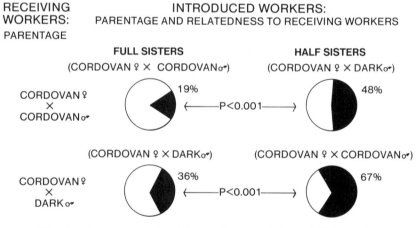

Figure 5.3 Results of test for discrimination between full and half sisters by worker honeybees. Workers were introduced individually into a group of 10 full-sister bees. Percentages shown denote the fraction of introduced workers which were bitten by at least one member in the receiving group. Introduced bees were bitten more frequently when placed amidst half sisters than when placed amidst full sisters. (Modified from Getz and Smith 1983.)

advantage that observations can be run blind. Evidently honeybees possess sufficient enzyme polymorphisms to do this with allozymes (Mestriner and Contel 1972, Martins et al. 1977, Contel et al. 1977).

Assuming that workers can distinguish full and half sibs, then for one patrilineal group to bias the colony's queen rearing to favor queens in their patriline would seem to require that the members of this group somehow dominate the other groups throughout the queen production process. Presumably such domination would require numerical superiority. Unfortunately, to date no good data exist which describe the detailed pattern of paternity of a colony's workers. As discussed in Chapter 3, it is clear that sperm of several males is used simultaneously, but just how many males are genetically represented among the female brood at any one time, and in what proportions they are represented remain key questions in honeybee biology. Is there any evidence at all that workers favor their full sisters in queen rearing? Probably the most that one can say at present is that if workers do select for queens of a certain patriline, they evidently do so before the queens reach adulthood. Although workers clearly will kill foreign queens entering their nest, they have not been observed to kill one of their own adult queens. When supernumerary queens occur in a colony, the workers leave it to the queens themselves to enter into the fights that establish which queens will survive to reproduce (Huber 1792, Allen 1956).

Life History Traits

Annual vs. perennial queens. In contrast to the females of most other bee species, which live for only one year, queen honeybees can live for several years, reproducing each year. I have seen labeled, three-year-old queens at the head of vigorous colonies, and beekeepers report that a significant fraction of queens can live for four or more years (Jean-Prost 1956, Bozina 1961). How has natural selection favored the unusually long lifespan of queen honeybees?

This question can be given focus through reference to the life history of a queen bee. We have already seen that queens are produced in the late spring or early summer of one year and that because of the need for extensive preparations for winter survival, a queen's first opportunity to produce reproductives (queens and males) generally does not come until the following spring, when the queen is approximately one year old. The question then arises, should the queen produce a single batch of reproductives in her lifetime, or should she attempt to produce several annual batches of reproductive offspring? (To simplify this discussion, we will leave males aside and think of queens reproducing primarily through the production of daughter queens.

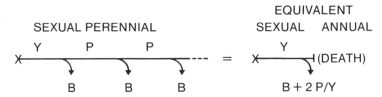

Figure 5.4   Reproduction of a sexual perennial compared with a more fecund sexual annual. An individual is born at X and then bears B or more offspring at the times shown by the arrows. P and Y denote the annual survival probabilities of parents and young. The two patterns are reproductively identical. (Modified from Horn 1978.)

This allows us to use a single currency—number of surviving queens—but does not influence the overall conclusions.) Given the honeybee's basic life history, in which one-year-old queens survive to future years by leaving the parental nest in a swarm (Seeley 1978, Winston 1980), the answer to the preceding question depends on how many colonies headed by daughter queens could be produced if the resources of the mother-queen's swarm were turned over to supporting one or more daughter queens. In general, a sexually reproducing organism should reproduce perennially if reproducing heavily in one year results in fewer total offspring than would be expected from a slower rate of reproduction spread over several years. Specifically, perennial reproduction by sexual organisms is favored when:

$$N < 2 \cdot (P/Y), \tag{5.1}$$

where N is the number of offspring made possible by the death of the parent, and P and Y represent the probabilities of parent and offspring survival to the following breeding season (Charnov and Schaffer 1973, Horn 1978) (Fig. 5.4). In particular, for the mother queen of a colony, the production of $2 \cdot (P/Y)$ colonies headed by daughter queens is genetically equivalent to her own continued survival and reproduction. Given the requirement expressed in equation 5.1, and assuming that the mother queen is still healthy, it is clearly to her advantage to avoid simply having a daughter queen replace her in the prime swarm. In this situation, $P = Y$ (the new colony created by a prime swarm probably has about the same survival probability whether headed by an old or a new queen), so $2 \cdot (P/Y) = 2$. But because only one offspring colony would be created by the mother queen's death, inequality 5.1 holds and thus it should be to the queen's advantage to live on. For a queen to benefit by being replaced while she is still a competent egg layer would require that the resources of the prime swarm could be partitioned among N daughter queens such that:

$$N \cdot Y(N) > 2 \cdot P, \tag{5.2}$$

where $Y(N)$ denotes the survival probability of each of the N colonies produced with the resources of the prime swarm. Although the precise relationship between swarm size (number of workers per swarm) and swarm survival has never been determined, I would guess that reducing swarm size by a factor of N leads to a reduction in the probability of swarm survival by a factor greater than N. If so, and given that $Y(1) = P$, then it is clear that inequality 5.2 cannot, as a rule, be fulfilled. Probably the only time it can be fulfilled is when, because of old age of a queen, P is very low. If, for example, $P < 1/2 \cdot Y(1)$, then a queen should be favored by natural selection if she allows herself to be replaced by a daughter queen.

What is the preference of a worker regarding her queen's lifespan? For workers, the critical difference between working for their mother queen or for a sister queen is the worker's relatedness to the reproductives produced by each type of queen. Assuming that workers control the ratio of sexual investment, but cannot bias the representation of the different patrilines among the queens that they rear, then the average relatedness of a worker to her mother-queen's offspring (r_{W-mQ_Y}) is given by equation 3.4:

$$r_{W-mQ_Y} = \frac{n^2 + 2n + 2}{4(n + 1)n}, \tag{3.4}$$

where n is the number of males mated with a queen. Following the same assumptions, the average relatedness of a worker to a sister-queen's offspring (r_{W-sQ_Y}) is precisely one-half of the average relatedness of a worker to the female offspring of her mother queen (equation 3.3), thus:

$$r_{W-sQ_Y} = \frac{1}{2} \cdot \frac{1}{2} \cdot \left(\frac{1}{2} + \frac{1}{n}\right)$$

$$= \frac{1}{8} + \frac{1}{4n}. \tag{5.3}$$

The ratio of these two average degrees of relatedness is

$$\frac{r_{W-mQ_Y}}{r_{W-sQ_Y}} = \frac{2 \cdot (n^2 + 2n + 2)}{(n + 1) \cdot (n + 2)}. \tag{5.4}$$

As n increases from 1 to 20, this ratio rises only slightly, from 1.67 to 1.91, so it can be approximated with fair confidence at about 1.8. Thus reproductives produced by a worker's mother queen are worth genetically about 1.8 times as much as ones produced by a sister queen. Therefore workers may be expected to retain their mother queen until her advanced age reduces a colony's overall fitness (ability to survive and produce reproductives) by a factor of about 1.8 relative to its fitness with a young, sister queen.

In summary, it appears that both the queen and the workers of a colony agree that the queen should be allowed to live until her egg-laying or some other property has declined to the point of roughly halving the colony's original ability to survive and reproduce. Although precise measurements of queen performance at the time of supersedure are lacking, it is generally understood that workers only rear a replacement for their mother queen once her egg-laying has declined dramatically (Butler 1974), and even then they do not kill their mother queen, but allow her to work together with the new queen. Presumably the workers hope to rear reproductives from their mother's eggs for as long as possible.

Level of reproductive effort. The prime resources of honeybees in their struggle for genetic survival are their bodies, their nest, and the food stored in their nest. When colonies undergo fissioning for reproduction in the female line, these resources become divided among several colonies, one of which is headed by the mother queen, the others by daughter queens reared in the parent colony (Fig. 3.7). How should a colony partition its resources? To answer this question it is useful to break it down into two slightly smaller questions:

(1) What is the optimal division of resources between the mother-queen colony and all of the daughter-queen colonies?

(2) Given a certain level of investment in colonies headed by daughter queens, what is the optimal number of daughter-queen colonies and how should resources be divided among these colonies?

These two questions form the themes of this and the next sections of this chapter.

Because in honeybees the mother queen leaves in the prime swarm, the first question can be recast more precisely as "How much of a parent colony's resources should be invested in the prime swarm and how much should be left in the parental nest after the prime swarm leaves?" What is observed is that roughly 70 percent of the parent colony's adult workers leave in the prime swarm (Martin 1963, Getz et al. 1982) and that these departing bees are probably a random sample of the adult workers in the nest when the prime swarm leaves, with the exceptions of very young (less than 3 days old) and very old (more than 30 days old) bees, which tend to remain in the parental nest (Butler 1940, Meyer 1956a). The mechanisms regulating this fissioning remain a mystery. What little is known is that a few workers trigger the fissioning by becoming buzz-runners (*Schwirrläuferin*), bees which excitedly press through their crowded nestmates, buzzing their wings every few seconds with bursts of 180–250 Hz vibrations. Periodically they stop, press against another worker, and generate a several-second pulse of 400–500 Hz wing buzzing (Esch 1967). The number of buzz-runners quickly multiplies until all the bees in a nest are scrambling about and eventually begin pouring out of the nest entrance (Lindauer 1955, Martin 1963). Over the next 20 minutes or so, most of the adult workers leave the nest, but circle about it in flight, staying within about 30 meters of the nest. Eventually a swarm cluster begins to form on a tree branch or similar support, the queen alights there, and gradually the flying workers begin to settle, some joining the swarm, others returning to the parent nest. Throughout this time, a process of equilibration seems to be in operation, with bees simultaneously landing and taking off at both the nest entrance and the swarm cluster. Gradually these comings and goings lessen, the air clears of flying bees, and finally two distinct groups of workers—the prime swarm and the parent-nest residue—have been formed.

Let us now view this process in terms of natural selection, a perspective which may both provide clues about the adaptive properties of the fissioning process and offer suggestions about the proximate mechanisms underlying this process. Most of the theory about the evolution of reproductive effort (Fisher 1930, Williams 1966a, 1966b, Schaffer 1974a, 1974b; reviewed by Stearns 1976) has been formulated from the perspective of the parent trying to maximize its lifetime production of offspring. In general, it is understood that the value of investment in immediate reproduction (as opposed to retaining resources for future reproduction) reflects an organism's demography. High rates of annual mortality combined with low yearly increases in fecundity favor immediate reproduction, whereas low annual mortality plus large yearly increases in fecundity favor the spreading of reproductive effort over several breeding seasons. These ideas are normally applied to a parent that must decide how to invest in offspring, but because the mother queen of a honeybee colony probably exerts only a minimal influence on how a colony's resources are partitioned between her survival and reproduction, we must turn instead to considering how natural selection has shaped the actions of workers.

To workers, their mother queen and any sister queens they rear are both essentially just channels through which workers can propagate their genes. Probably the primary difference to a worker between a mother queen and a sister queen lies in their effectiveness in transmitting a worker's genes into the next generation. Assuming that queens are inseminated by about 10 males, that all males are represented equally and randomly among a queen's female offspring, and that workers cannot bias the representation of the different patrilines among the new queens, then the average degrees of relatedness of a worker to her mother's and sister's offspring are about 0.28 (r_{W-mQ_Y}) and 0.15 (r_{W-sQ_Y}), respectively (see equations 3.4 and 5.3).

The simplest choice situation faced by workers arises when no afterswarms leave the parental nest, and therefore all the workers not leaving in the prime swarm stay and help the sister queen that inherits the parental nest. Bulmer

(1983) has explored theoretically how the workers should divide themselves between the two queens. If the parent colony contains W workers, of which x_M go with the mother queen and x_S stay with the sister queen, and $f_M(x_M)$ and $f_S(x_S)$ are the probabilities of these two queen's colonies surviving to maturity (Fig. 5.5), then during colony fissioning workers are expected to distribute themselves between the two colonies such that:

$$r_{W-mQ_Y} f'_M(\hat{x}_M) = r_{W-sQ_Y} f'_S(\hat{x}_S), \qquad (5.5)$$

where $f'_M(x_M)$ and $f'_S(x_S)$ are the derivatives of the survival functions $f_M(x_M)$ and $f_S(x_S)$ and $\hat{x}_M$ and $\hat{x}_S$ are equilibrium values. (Note that the two functions are not identical, primarily because the sister-queen colony inherits a complete nest while the mother-queen colony must construct a new nest.) Under these conditions there is an equilibrium; if a worker should shift from the mother-queen colony to the sister-queen colony, the worker would lose as much in inclusive fitness through the mother queen as she gains through the sister queen. In other words, under the equilibrium conditions each worker is expected to be indifferent as to which colony she should join (Fig. 5.5). Depending on the shapes of the survival functions f_M and f_S, there may be many or few pairs of x_M and x_S satisfying equation 5.5. The genetic payoff (G) to the workers of each such pair is given by:

$$G = r_{W-mQ_Y} f_M(\hat{x}_M) + r_{W-sQ_Y} f_S(\hat{x}_S), \qquad (5.6)$$

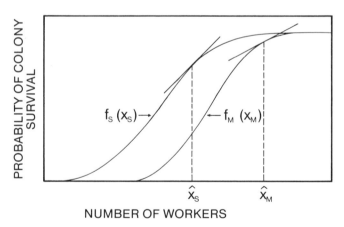

Figure 5.5 Graphical illustration of the relationship between the survival function for the mother queen ($f_M(x_M)$) and sister queen ($f_S(x_S)$) colonies, and equilibrium values of the number of workers leaving with the mother queen ($\hat{x}_M$) or staying with the sister queen ($\hat{x}_S$). The equilibrium values shown are only one of many possible pairs satisfying equation 5.5, where $r_{W-mQ_Y} = 0.28$ and $r_{W-sQ_Y} = 0.15$.

and it seems reasonable to conjecture that workers will evolve to achieve the equilibrium values $(\hat{x}_M{}^*, \hat{x}_S{}^*)$ which yield the highest payoff.

The critical test of these ideas will come from empirically determining the shapes of the survival functions f_M and f_S, combining this with information about the relevant degrees of relatedness, and, finally, comparing the predicted values of $\hat{x}_M{}^*$ and $\hat{x}_S{}^*$ with what is actually observed.

Number of daughter colonies. Two fundamental traits of any species' reproductive process are the number of offspring produced per parent and the level of investment per offspring. Because the resources for reproduction are always finite, there must be a tradeoff between these two variables, with individuals rearing either a few expensive young, numerous inexpensive ones, or some intermediate combination of number and cost. An intriguing feature of honeybee reproduction is their contrasting investment patterns for the two sexes. Each summer, a typical colony produces several thousand male reproductive units (drones), each one weighing only about 220 milligrams, but only one to four female reproductive units (assemblages of a virgin queen plus several thousand workers), each of which typically weighs more than a kilogram, thus some 5000 times the weight of a drone (Mitchell 1970, Fell et al. 1977, Winston 1980). As was explained in Chapter 4, this massive investment per queen evidently arose in the tropics, in response to intense predation on colonies, and has been maintained in temperate regions, where it is crucial to the winter survival of queens. In stark contrast, reproductive success through males does not depend on long-term survival, but instead on effectiveness in finding and mating with queens. These circumstances evidently favor the production of numerous, inexpensive males.

For the remainder of this section, I will focus on the question of the number of female reproductive units produced by a colony each summer, that is, female brood size. Because these female units are each an assemblage of physically independent subunits—a queen, several thousand workers, and sometimes nest materials and food stores—they represent a reproductive system which is unusually open to experimental manipulation. Thus honeybees provide an especially attractive organism for testing hypotheses about the evolution of brood size.

Current hypotheses on the adaptive significance of an organism's brood size (reviewed by Stearns 1976) are largely elaborations on David Lack's (1954, 1966) idea that animals are selected to have the brood size which is most productive—that is, results in the maximum number of young surviving to maturity. The principal modification of this idea has concerned the counting of surviving young. Lack counted survivors per year, but in general it seems more appropriate to count surviving young over the lifetime of a parent (Charnov and Krebs 1973, Williams 1966b). Since parent survival is influenced by brood size, with parents rearing the largest possible number of young in one year having strongly reduced their chances of surviving to reproduce another year, the optimal brood size for a single breeding season is generally expected to be somewhat smaller than the largest possible brood. However, applying such ideas to honeybees and other social insects is complicated by the prominent role of the offspring (the workers) in shaping the reproductive patterns. Thus, instead of casting our hypotheses on the evolution of female brood size from the perspective of a parent selected to maximize its reproductive success, in social insects it is more appropriate to analyze the perspective of workers selected to maximize their inclusive fitness.

A honeybee colony's investment in female reproduction consists of all the resources left behind in the parental nest following the departure of the mother-queen's swarm. These resources can be kept intact as a single colony, or can be divided up among several daughter queens by producing afterswarms. The only quantitative description of what unfolds in temperate-zone colonies after the prime swarm leaves is that of Winston (1980). As is shown in Table 5.2, nearly all colonies generated one or more afterswarms as well as a prime swarm. The average female brood size was 2.5 colonies: 1.5 afterswarm colonies plus 1 colony in the parental nest. Although the internal events in a colony following the mother queen's departure have only rarely been observed closely (Huber 1792, Allen 1956, Simpson and Cherry 1969), the available observations suggest that workers are capable of closely regulating the further fissioning of their colony. Most importantly, the workers evidently can control the interactions between rival virgin queens. Their control techniques include postponing a virgin queen's emergence from her cell by not removing the

Table 5.2

Number and sizes of swarms produced by eight honeybee colonies in one summer. (From Winston 1980.)

Colony	Prime swarm	1st Afterswarm	2nd Afterswarm	3rd Afterswarm
		Number of workers in		
1	11,676	11,076	?	—
2	13,529	6,091	4,086	—
3	?	—	—	—
4	21,818	10,608	—	—
5	14,824	13,778	3,765	4,296
6	17,260	13,127	?	—
7	12,978	14,625	—	—
8	20,143	11,458	—	—

? indicates a swarm which was not captured and measured.

tough wax and cocoon fibers on the tip of her cell, chivying already emerged queens away from unopened queen cells, keeping two emerged queens apart by pinning them in place, and forcing queens to leave the nest in afterswarms. Overall, it appears that the virgin queens have relatively little control over whether they will leave in an afterswarm or inherit the parental nest.

Assuming that workers do exercise powerful control over the fates of virgin queens, it seems likely that an important goal of an emerged queen is demonstrating that she promises to be a vigorous queen, one in which the workers should invest their resources. This message may be communicated in the loud piping sounds of queens (fundamental frequency about 435 Hz, Hansson 1945), which they produce by pressing the thorax against the comb and vibrating their flight muscles but without spreading their wings (Simpson 1964). Although plainly audible to humans as airborne sounds, worker bees detect them as substrate-borne vibrations (Little 1962). The immediate effect of these sounds on neighboring workers is that they stand motionless on the combs; the long-term effect may be to inhibit the workers from releasing additional queens until the currently piping queen has departed in a swarm. Consistent with this interpretation are the observations that a mother queen only began piping after the cells of her daughter queens were sealed, and that when virgin queens emerged from these cells, they piped only if guarded, unopened cells remained in the nest. Furthermore, the piping of queens, which can occur as frequently as every minute or so, occurs primarily when a queen is near a queen cell, suggesting that queen cells provide a special stimulus to piping (Allen 1956).

Given that workers can closely regulate the number of afterswarms produced by a colony, how can we understand their decision regarding the number of afterswarms? If we assume that the decision is made irrespective of the kin relationships between the workers and the various virgins who will take over the new swarms, then one can hypothesize that selection will favor the fissioning pattern which yields the largest number of offspring colonies surviving to maturity. The general solution to this problem is shown in Figure 5.6. The basic decision is this: a few large offspring colonies or many small ones? The answer depends on the probability of colony survival, which probably depends greatly on the initial size of colonies. Although the precise shape of the curve of colony survival versus number of colonies is unknown, and therefore a precise prediction of the best number of offspring colonies cannot be made now, this hypothesis predicts in general that there exists an optimal number of colonies which is intermediate between so few that resources are underutilized, and so many that the resources are spread too thin among the colonies. Experiments by Simpson and Cherry (1969) provide some support for this hypothesis. Their studies indicate that workers respond to high colony strength, as indicated by either congestion or number of bees in the nest, by

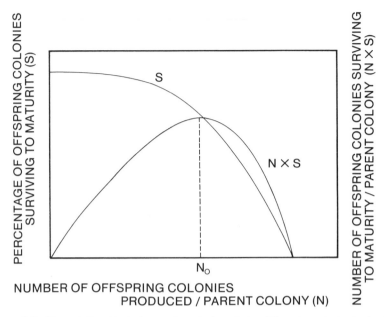

Figure 5.6   The relationship between female brood size (*N*) and survival rate for the first year of life (*S*). The resulting $N \cdot S$ or production curve has a single maximum at N_0. This is the brood size favored by natural selection.

confining virgin queens to their cells instead of allowing them to emerge freely. Because such queen confinement, followed by sequential release of one or more queens, sets the stage for afterswarming, it appears that colonies adjust their frequency of afterswarming according to colony strength.

An alternative hypothesis for the number of afterswarms assumes that workers can discriminate full sisters from half sisters. This hypothesis employs the idea of Bulmer (1983), described above, in which each worker's choice between staying in the parental nest or leaving in a swarm depends upon her relatedness to the future offspring of each colony's queen, and the survival functions of each colony. It seems that the most likely question to be decided after the mother queen leaves would be whether to stay and help a half-sister queen or to leave with a swarm headed by a full-sister queen. Given that a worker's relatedness to a full sister's future offspring ($r = 0.375$) differs greatly from that to a half sister's future offspring ($r = 0.125$), workers may be strongly selected to produce afterswarms. If the net gain (equation 5.6) favors afterswarming at all, then by equation 5.5 workers will be selected to leave in an afterswarm so long as the swarm's (the full-sister queen's) probability of survival to maturity is greater than one-third that of the parent nest colony (containing a half-sister queen). The very small second and third afterswarms observed by Winston (1980) (see Table 5.2), which seem unlikely to enhance a colony's total production of queens surviving to maturity, suggest that kinship effects do influence the frequency of afterswarming. The observations of Getz et al. (1982), that swarming colonies show a moderate tendency to divide into patrilineal groups, also support the kinship-factor hypothesis for number of afterswarms. However, the issue is far from settled.

Mating System

Precise knowledge of the honeybee's mating system is vital to understanding this insect's social behavior. This stark statement reflects the fact that an animal's mating behavior powerfully shapes such major parameters of social evolution as the intensity of inbreeding, the level of local mate competition, and the degree of relatedness among nestmates. Until the late 1950's and early 1960's, progress toward understanding honeybee mating was slow, hindered by the fact that queens and drones mate high in the air, far from their nests. At this time, however, came two major discoveries—first, that matings occur in special congregation sites (Jean-Prost 1958), and second, that queen substance ((E)–9–oxo–2–decenoic acid) functions as a sex attractant pheromone (Gary 1962). These findings provided essential insights into how one could study the honeybee mating system, and so begin to perceive the genetical structure of honeybee populations.

It is now known that when virgin queens and drones fly out of their nests on mating flights, they do not wander aimlessly over the countryside, relying upon chance encounters with members of the opposite sex, but instead fly toward specific mating sites, called "drone congregation areas" (Ruttner and Ruttner 1963, Zmarlicki and Morse 1963). Drones leave their nests starting about 13.00 hours, about one hour before virgin queens fly out, and therefore are already at the congregation areas waiting for the queens by the time they arrive (Ruttner 1962, Ruttner and Ruttner 1965). The physical basis for drone congregation areas remains an intriguing mystery. It is clear that the same sites are used year after year even though all drones die at the end of the summer and so cannot lead the next year's crop of drones to the mating areas. In the Austrian Alps, for example, one set of especially well-documented drone congregation areas has persisted unchanged for at least 12 years (Ruttner 1976, Ruttner and Ruttner 1972) and numerous other areas have been monitored without change for shorter periods (Jean-Prost 1958, Ruttner and Ruttner 1966, Böttcher 1975). It is also clear that these areas, though they range in size from about 50 to 200 meters in diameter, have remarkably distinct boundaries, so that displacing a queen laterally 30 meters or so can reduce by tenfold the number of drones hovering around the queen (Ruttner and

Ruttner 1965). Circumstantial evidence suggests that queens and drones orient over large distances to the congregation areas by flying toward depression points on the horizon line where there is also high contrast between the earth and sky (Ruttner and Ruttner 1966, 1972), but this observation still leaves unsolved the puzzle of what defines the precise boundaries of mating areas. Drone congregation areas have been found over valleys, mountain ridges, open fields, and forests, thus there is no obvious common property in the underlying terrain or vegetation.

Further important mysteries about these mating sites concern their density and the spatial distribution of a colony's reproductives across the surrounding congregation areas. One intensive search near Erlangen, West Germany, revealed five drone congregation areas within a circular area of radius one kilometer (Böttcher 1975); studies near Lunz-am-See, Austria, suggest a comparable density of one congregation area per square kilometer (Fig. 5.7). It is clear that queens and drones can visit very distant drone congregation areas. By labeling the drones from colonies scattered throughout a valley in the Austrian Alps, and capturing drones at a congregation area, it was found that drones will fly 7 or more kilometers to reach a congregation area, flying up and over 1000-meter high mountains if necessary to reach a mating site (Ruttner 1976, Ruttner and Ruttner 1966, 1972). This figure for the flight range of drones, combined with the observation that queens frequently mate with genetically marked drones from hives 12 or more kilometers away, indicates that queens can also travel several kilometers from their nests before mating (Peer 1957, Peer and Farrar 1956, Ruttner and Ruttner 1972). Although these large distances probably represent the maximum flight ranges of reproductives, not their typical flight distances, various studies also indicate that queens and drones rarely mate at the congregation areas nearest their nests, with queens mating on average 2 to 3 kilometers from their nests (Ruttner and Ruttner 1972, Böttcher 1975). Certainly these numbers suggest that outbreeding is the general rule for honeybee populations. Assuming that the density of colonies (queens) in nature is one to two colonies per square kilometer (Taber 1980, Visscher and Seeley 1982), that colonies within a circle of diameter 12 kilometers interbreed, and that each queen mates on average with 10 drones ($N_M = 10 \cdot N_F$), then applying Wright's (1933) formula for the effective population size (N_e) of haplodiploid organisms:

$$N_e = \frac{9 \cdot N_M \cdot N_F}{4 \cdot N_M + 2 \cdot N_F} \tag{5.7}$$

yields an estimate of 240–480 individuals, which roughly agrees with the two estimates of Yokoyama and Nei (1979): 428–1093 individuals. Evidently the honeybee's mating system gives this insect reasonably large effective breeding populations despite the low density of its colonies.

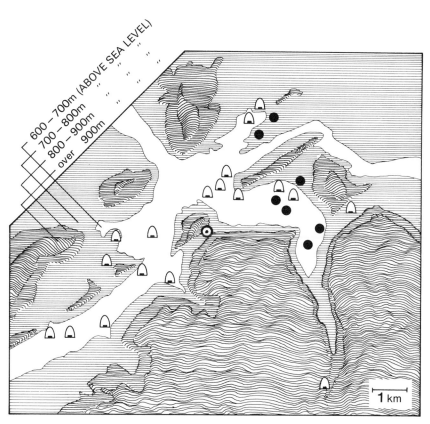

Figure 5.7 Locations of drone congregation areas (circles) and apiaries (skep symbols) in the mountains around Lunz-am-See, Austria. Mark-and-recapture studies revealed that drones from all the apiaries indicated were visiting the drone congregation area denoted by the open circle in the center of the map. (Modified from Ruttner and Ruttner 1966.)

The honeybee's mating process is understood in far finer detail starting with the time at which a queen enters a drone congregation area. In essence, each queen is polyandrous but each drone is monogamous. This unfolds as follows. When a queen flies into a mating area, an aerial trail of (E)–9–oxo–2–decenoic acid streams out behind her, attracting males 30 or more meters downwind (Gary 1962). As many as 100 or more drones then converge on the queen, all scrambling to make contact with her, and so creating a comet-tail formation tracking just behind the flying queen. Upon contacting the queen, it takes a drone a scant 2 to 4 seconds to mount and inseminate her, with mating culminating in an explosive snap as the drone's genitalia evert, semen is shot in the queen, and the drone, now paralyzed, falls back and away from the queen (Gary 1963, Koeniger et al. 1979). Immediately there-after, another drone can mount and inseminate the queen, and so the process can be quickly repeated, with each queen receiving sperm from 10 or so drones (see Chapter 3). The queen readily limits the number of matings by simply closing her sting chamber. An experimental analysis with wooden, dummy queens found that an opening of diameter 3.2 to 4.8 mm (the size of a queen's open sting chamber) on the posterior end of the queen is essential for mating. Without it, drones will mount, but they do not fully evert their genitalia or inject semen (Gary and Marston 1971).

The total volume of semen acquired by a queen during her visit to a drone congregation area greatly exceeds what she will store for use thoughout the rest of her life. Each drone injects about 11 million sperm into a queen and the total number of sperm received on a mating flight is some 87 million. Yet a queen typically stores only about 5 million sperm in her spermatheca (Woyke 1960). Although it is not clear just how randomly a queen samples from among the 87 million sperm in storing away her 5 million sperm, the way in which the sperm are processed before being stored suggests that a great deal of mixing of the different drones' gametes can occur. During mating, the sperm are received into the queen's lateral oviducts. Upon returning to her nest, the queen forces the sperm into her vagina by muscular contractions. A valve-like fold stops much of the semen from flowing back out, and instead directs the sperm into the spermathecal duct and thence into the spermatheca. The excess semen pushes out past the vagina fold and out of the queen in the form of thin threads which are removed by the workers (reviewed by Snodgrass 1956, Ruttner 1956, 1968d).

6 Nest Building

Selecting a Nest Site

In many animal species, especially certain birds, rodents, and social insects, individuals carefully choose a particular microhabitat in which to build nests and rear offspring (Lack 1968, von Frisch 1974). Such behavior has major adaptive significance since it can help ensure refuge from harsh physical conditions, an adequate food supply, and protection from predators. Honeybees provide an example of an extremely sophisticated process of nest-site selection. No fewer than seven distinct properties of a potential home site—including cavity volume, entrance size, distance from the parent nest, and presence of combs from an earlier colony—are independently assessed to produce an overall judgment of a site's quality. Nest-site selection by honeybees is further intriguing because it is a social process, one which involves several hundred individuals simultaneously scouting the environment in a coordinated hunt for the best available dwelling place. This massive search operation usually reveals twenty or more potential home sites, only one of which is finally chosen for habitation. Although it is difficult to quantify the matter, it seems certain that the effectiveness of these group searches goes far beyond what any solitary insect can achieve.

The honeybee's process of househunting arises in the course of colony reproduction but begins even before a swarm has left the parent nest (Lindauer 1955). The start comes when a few hundred of a colony's oldest bees, its foragers, cease collecting food and turn instead to scouting for new living quarters. This requires a radical switch in behavior. No longer do these bees probe bright-colored, sweet-scented sources of nectar and pollen; instead they investigate dark places—knotholes, fissures in rock, cracks in tree limbs, and gaps among roots—always seeking a small cave suitable for enclosing a honeybee nest. Upon discovering such a site, a scout spends nearly an hour examining it closely. Her inspection consists of a series of roughly minute-long excursions inside the cavity alternating with trips outside. While outside, the scout scurries over the nest structure and performs slow, hovering flights all around the nest site, apparently conducting a detailed visual inspection of the structure and surrounding objects. While inside, the bee scrambles over the interior surfaces, at first not venturing far inside the cavity, but with

increasing experience penetrating deeper and deeper into the remote corners
of the hollow (Fig. 6.1). When her examination is complete, a scout will
have walked 50 meters or more around the inside of the cavity and so will
have covered all of its inner surfaces. Experiments with cylindrical nestboxes
whose walls could be rotated (thus placing a scout bee on a treadmill when
she is inside the nestbox) have demonstrated that scout bees assemble a
perception of a cavity's volume from the amount of walking required to
circumnavigate the space (Seeley 1977).

The long duration of nest-site inspections suggests that scouts assess multiple properties of a site to judge its suitability, although a clear picture of
the astonishing complexity of the evaluation process has only recently begun
to emerge through experimental studies of the bees' preferences in a home
site (Seeley and Morse 1977, 1978, Jaycox and Parise 1980, 1981, Gould
1982, Rinderer et al. 1982). The basic design of these experiments involves
providing swarms with a choice among nestboxes differing in a certain property, such as volume or entrance area, and observing which boxes the swarms
routinely occupy. Such studies show that Italian honeybees (*Apis mellifera
ligustica*) favor a nest cavity whose (1) volume falls in the range of 15 to 80

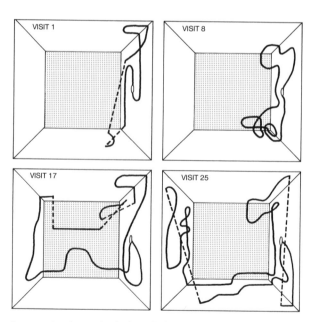

Figure 6.1 A scout bee's method of exploring a potential nest cavity is indicated by
tracings of what a single scout did on 4 out of 25 visits during her initial inspection of
a possible home site. Where the line is solid, the bee was walking; where it is broken
she was flying. The hole on the right is the entrance to the cavity. (From Seeley 1982b.)

liters, and whose (2) entrance faces south, (3) is smaller than 75 cm², and (4) is positioned near floor level but (5) is at least several meters above the ground. The ideal nest site for this race of bees also (6) lies between 100 and 400 meters from the parent nest and (7) comes equipped with a complete set of beeswax combs, built by a previous colony. Without doubt, other nest-site properties besides these seven also influence a site's attractiveness to bees.

The ecological significance of each of these home-site preferences remains to be experimentally tested, for example, by comparing the survivorship patterns of colonies inhabiting hives which differ in volume, entrance size, or some other variable. Nevertheless, several benefits to honeybees from applying their nest-site preferences seem clear. First, it is probably critical that honeybees avoid cavities smaller than about 15 liters since, as we have seen (Chapter 4), a colony consumes at least 10 kilograms of honey over the winter. This mass of food requires a minimum of about 15 liters of storage space. Successful overwintering is probably also facilitated by having a proper nest entrance. A small entrance helps insulate the warm winter cluster from cold temperatures outside, and an entrance at the bottom of the nest cavity rather than at the top may help minimize heat loss from the nest by convection currents. A south-facing entrance probably also helps by providing a solar-heated porch from which bees can make their critical cleansing flights to eliminate accumulated body wastes on mild days in mid-winter. A sunny entrance in winter also reduces the possibility that the entrance opening will become plugged by ice and snow (Szabo 1983a). Moving into a site already furnished with combs must allow nascent colonies to shunt energy which would otherwise be allocated for the building of combs into brood rearing or food storage and thus boost a colony's odds of winter survival. Szabo (1983b) observed that swarms installed in hives containing a full set of combs amassed nearly twice as much honey across a summer season as swarms placed in empty hives (81 and 43 kilograms, on average, respectively). Finally, care in choosing lodgings almost certainly reduces the threat of predation to a colony. High entrances are inaccessible to predators that cannot fly or climb trees, and are harder to find than entrances near the ground. Even if the nest should be discovered, a small entrance effectively shrinks a colony's defense perimeter to the scale of a few centimeters.

In time, a scout bee which has discovered a possible home site completes her inspection and returns to her colony. There she spreads the news of her discovery among fellow scout bees by performing waggle dances, which code the distance and direction of her find (see Chapter 7). At the same time other successful scouts will also begin advertising their own troves. Overall, some two dozen (24 ± 8, range 13–34) potential nest sites will be reported by a colony's scouts (Lindauer 1955). This wealth of alternatives helps ensure that colonies locate the best possible site, but it also demands that they be able to decide which is best. The scouts will need several days of steady effort to sort through the possibilities and identify the single best dwelling place. Although their investigations begin before the swarm emerges from the parent nest, a final decision is generally not reached until after the swarm has flown from the parent nest and bivouacked in a temporary cluster on a tree branch nearby (see Fig. 5.1).

The heart of the decision-making process is the ability of scout bees to switch their preferences among nest sites. Thus although one scout may discover a particular site, and initially advertise it because it is the best site she knows, if a second scout finds a better site, the first scout will eventually shift her allegiance to the superior alternative. Such switching results from each scout coding the quality of her site in the vigor of her recruitment dances. Exceptional sites are represented by lively dances that last for half an hour or more, whereas mediocre sites merit only sluggish, seemingly unenthusiastic dances. When a scout which has been performing leaden dances encounters one dancing energetically, she reads this scout's dances and flies off to inspect the corresponding site. If her inspection reveals that it is indeed superior, she begins advertising it in her own dances on the swarm (Fig. 6.2). Thus one by one the scouts gradually transfer their attention from deficient sites to ever better ones, and so ultimately reach a consensus about which dwelling place is best.

When an agreement has been reached, only about 500 scout bees, approximately five percent of a swarm's population, know the precise location of their new abode, and thus the problem remains of spreading this information among all members of the swarm. The scouts have communicated among themselves the locations of many home sites by means of waggle dances. But rather than employing this communication system on a larger scale to inform the nonscouts in their swarm, the scouts instead simply trigger the swarm to launch into flight with buzzing runs (Esch 1967; see Chapter 5) and then they pilot the 10-meter-diameter cloud of flying bees in the proper direction (Seeley et al. 1979). Precisely how the small crew of scouts guide their thousands of sisters remains unclear. Possibly they streak through the swarm cloud in the direction of the nest site and so literally point the way (Lindauer 1955). Once the swarm reaches its destination, the scouts somehow signal it to stop. Then they quickly drop out of the cloud of circling bees, alight at the nest site's small entrance opening, and release assembly pheromone from the Nasonov gland at the tip of the abdomen to pinpoint the nest entrance. Within minutes the other bees begin streaming into their new home. Thus the nest-site selection process, involving hundreds of scouts, dozens of alternative dwelling places, and probably several thousand independent decisions by the scout bees, draws to a close.

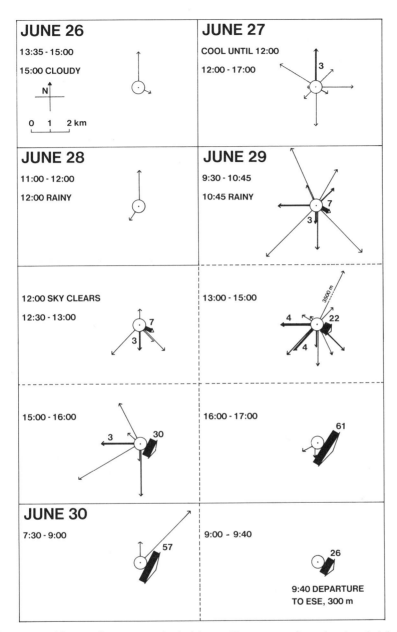

JUNE 26

13:35 - 15:00

15:00 CLOUDY

N

0　1　2 km

JUNE 27

COOL UNTIL 12:00

12:00 - 17:00

3

JUNE 28

11:00 - 12:00

12:00 RAINY

JUNE 29

9:30 - 10:45

10:45 RAINY

7

3

12:00 SKY CLEARS

12:30 - 13:00

7

3

13:00 - 15:00

3500 m

4　　22

4

15:00 - 16:00

3　30

16:00 - 17:00

61

JUNE 30

7:30 - 9:00

57

9:00 - 9:40

26

9:40 DEPARTURE
TO ESE, 300 m

Figure 6.2   History of one swarm's decision-making process from the time it left its old nest (at 13:35 hours on June 26) until it moved into the new nest site (at 09:40 hours on June 30). Each circle represents the location of the swarm; the arrows indicate the distance and direction of potential nest sites. The thickness of each arrow denotes the number of new bees dancing for a site over the time interval shown. (Modified from Lindauer 1955.)

Comb Construction

Finding a proper home site is only the first of several major hurdles which newly founded colonies must surmount if they are to survive the following winter. Other obstacles include rearing new workers that can live through winter, and gathering the twenty or so kilograms of honey needed for winter provisions. Before even starting on these problems, however, a colony must build a nest of beeswax combs (Fig. 6.3). The cells in these combs function both as cradles for brood and as storage containers for pollen and honey.

Honeybee nests are expensive. When completed, a typical nest will contain some 100,000 cells arranged in combs whose total surface area is about 2.5 square meters (Seeley and Morse 1976). Building this impressive edifice requires over 1200 grams of beeswax, which is a complex blend of straight-chain monohydric alcohols (C_{24}–C_{36}) esterified with straight-chain carboxylic acids and hydroxycarboxylic acids (C_{16}–C_{36}), mixed with various straight-chain alkanes (C_{21}–C_{33}) (Callow 1963). The difficulty of completing this massive construction is underscored by the fact that swarms must assemble nearly all their building materials from scratch; a swarm brings very few resources for nest construction from the parent colony. On average, a worker bee in a swarm contains about 35 milligrams of a 65 percent sugar solution (Combs 1972), and thus the 12,000 bees in a typical swarm (Fell et al. 1977)

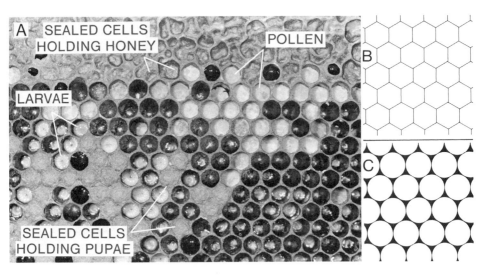

Figure 6.3 (A) Comb on the edge of the broodnest showing cells containing brood and food. (B) and (C) Illustration of the economy in wax achieved by building hexagonal cells.

altogether hold an energy reserve of only some 275 grams of sugar. Given that the weight-to-weight efficiency of beeswax synthesis from sugar is at best about 0.20 (Horstmann 1965, Weiss 1965), and that one gram of wax yields about 20 cm² of comb surface (Ribbands 1953, Root 1980), the complete conversion of an average swarm's sugar supply into comb would produce at most about 1100 cm² of comb surface—scarcely 4 percent of a complete nest's comb. The overall cost of nest building can also be reckoned in terms of the cost of overwintering. Given that honey is an approximately 80 percent sugar solution and the aforementioned efficiency in converting sugar to wax, then the 1200 grams of wax in a typical nest is energetically equivalent to about 7.5 kilograms of honey, or about one-third of the food consumed by a colony over winter.

These numbers demonstrate that the cost of nest construction constitutes a large fraction of a colony's energy budget during its first year. This suggests that colonies can benefit greatly from energy conservation during nest building. In fact, a colony's very survival may depend on such frugality. In central New York State, for example, only 24 percent of the newly established (wild) colonies survive to the following spring. The vast majority perish in midwinter when their honey stores become exhausted (Seeley 1978).

Nest building commences as soon as a swarm enters its new nest cavity. First the bees gnaw away any loose, crumbly punkwood from the cavity's ceiling, thus preparing a solid surface from which to suspend the new combs. Next the bees assemble themselves into a cluster of interconnected chains of bees suspended from the top of the cavity. For the next 24 hours or so, nearly all of the bees in the swarm, with the exception of the foragers, hang here almost motionless, all the while secreting tiny scales of wax from glands on the underside of their abdomens. Comb formation starts when individual bees with well-developed wax scales disconnect themselves from their sisters, climb upward through the braids of bees, and deposit their wax after chewing each wax scale to mix it with a mandibular gland secretion which renders the wax more plastic. Initially these deposits produce simply small piles of wax, but eventually the piles merge into a ridge of wax several millimeters long and 2 to 4 mm high. At this point the sculpting of cells begins (Huber 1814, Darwin 1859, Darchen 1968). First a cavity the width of a worker cell is excavated in one side of the wax ridge, with the bees depositing the removed wax along the sides of the hole. This is repeated on the other side of the ridge, but here two cells are dug, with the center of the first cell between the opposite two. Next the raised edges are converted to linear prominences, the future bases of each cell's walls, with adjacent walls laid out at an angle of 120° to give the cell a hexagonal cross-section from an early stage. As additional wax is deposited, the bases of adjacent cells take shape at the appropriate distances from pre-existing cells, and the walls of these earlier cells are raised by adding rough particles of wax to the top of each wall and then shaving this down on both sides to form a thin, smooth, plane of wax in the middle. The cut-away wax is then piled up, together with fresh wax, on the top of the ridge and the process is repeated. Thus a thin wall grows steadily upward, always crowned by a broad coping.

Throughout this process of comb construction there runs a theme of economy in wax. The most conspicuous expression of this is the honeybee's basic cell shape: a right hexagonal prism capped on the inner end by a trihedral pyramid. Because honeybee cells were originally round in cross-section, as they still are for most species of long-tongued, advanced bees (Michener 1964), one can view a honeybee comb as, in essence, an assemblage of co-equal cylinders, of circular cross-section, compressed into hexagonal prisms (Thompson 1942). Although a hexagon of a given area possesses a 5 percent longer perimeter relative to a circle of the same area, because each hexagonal cell in a comb shares its walls with other cells whereas circular cells do not, for cells of a given volume and given wall thickness, a comb of hexagonal cells requires only about 52 percent of the wax needed to build the comb of circular cells (Fig. 6.3). For example, the wall-to-wall distance of honeybee worker cells is about 5.25 mm (Darchen 1968). Hexagons of this size have a perimeter of 18.19 mm and an area of 23.87 mm^2. Circles of this size have a perimeter of 17.32 mm. But because each cell wall in a hexagonal-cell comb is shared by two cells, the effective perimeter per hexagonal cell is 18.19 mm/2 = 9.09 mm, and 9.09/17.32 = 0.52.

Several other features of the comb construction process, besides the hexagonal design, also contribute to the honeybee's economy in use of wax. One is the bee's high skill in planing down the wax partitions between cells, thereby leaving the walls and bases of cells only 0.073 $\pm$ 0.008 and 0.176 $\pm$ 0.028 mm thick, respectively (Fig. 6.4). A worker bee's antennae evidently play a critical role in the process of measuring wall thickness. When the six distal segments of bees' antennae were amputated, their building practices were grossly disrupted; some cells were built with holes gnawed in their walls while other cells received walls 118 percent thicker, on average, than normal (Martin and Lindauer 1966). Because the temperature and composition of their building materials are constant, and because the shape of their cells is also uniform, it seems likely that worker bees can judge the thickness of a cell's walls by pressing on them with their mandibles and noting the elastic resiliency of the substrate with their antennae. Bees further minimize their colony's need for wax by constantly recycling old wax. Whenever a bee emerges as an adult, the cappings of her brood cell are carefully bitten off by the emerging bee or nearby nurse bees and are then stuck to the cell's rim for use again later. Likewise, queen cells are built from bits of wax cut away from adjoining worker cells, and vacated queen cells are torn down to free

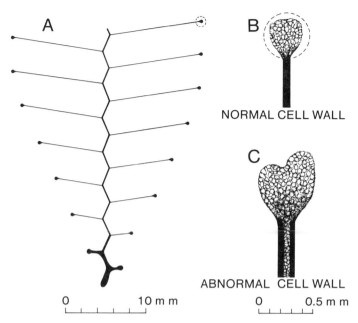

A

B

NORMAL CELL WALL

C

ABNORMAL CELL WALL

0   10 m m   0   0.5 m m

Figure 6.4 (A) Cross-section through a honeybee comb showing the way bees pre-cisely plane down cell walls during construction to minimize the cost of comb building. (B) and (C) Cross-sections of the outer margins of cells constructed by normal bees (B) and bees from which the 6 distal segments of their antennae were amputated (C). Normally, loose bits of wax are packed together at the tip of the cell wall while the rest of the wall is a smooth, thin layer of wax. Bees with antennal operations build costly, triple-layered cell walls. (Modified from Martin and Lindauer 1966.)

wax for other purposes. Even the mechanisms regulating wax secretion by a colony's worker force appear designed to hold a colony's wax production to a bare minimum. Aside from the time of initial nest building, the only time wax is again produced in abundance is when a colony lacks storage space for a large crop of honey. This tight linkage between the need for wax and its production evidently results from having the same group of bees—the food storage or caste III bees (see Chapter 3) (Rösch 1927, Seeley 1982a)—responsible for both wax production and honey storage. When these bees lack empty cells in which to deposit fresh nectar, they must function as living food reservoirs, their honey stomachs filled with nectar. If this situation persists for more than a few hours, their wax glands begin secreting wax, and this wax production, in turn, triggers a timely bout of comb building (Ribbands 1952, Butler 1974).

7

Food Collection

Introduction

Foraging by honeybees is a social enterprise, one in which the 10,000 or so foragers in a colony work together closely in finding and exploiting rich sources of nectar and pollen. One key to understanding this system of group foraging is recognizing that it is designed to achieve high efficiency for the colony as a whole, not for each individual forager. This curious fact is most apparent when a forager performs behaviors which reduce her personal rate of food collection but which boost her nestmates' intake rates and so enhance the colony's overall foraging efficiency. The clearest example of such a behavior comes when a bee discovers a highly profitable patch of flowers and recruits her sisters to her rich find. This recruitment depresses the first bee's rate of food retrieval, both because it consumes time she could spend collecting and because it ensures that she will soon have to share her patch with more and more bees. Simultaneously, however, this recruitment elevates her colony's overall feeding rate because it directs previously unemployed foragers to a profitable work site. A second example of individual efficiency being sacrificed for colonial efficiency arises in the search for new forage patches. When individual foragers need to find a new work site, most do so by following recruitment dances (see below) but a small minority simply fly out and search independently (Lindauer 1952, Seeley 1983). These are scout bees. It seems likely that most scouts will be less successful foragers than their sisters which follow recruitment dances; scouts search randomly for patches whereas recruits are guided to top-quality patches. However, a few scouts make the all-important discoveries of major new food sources, ones capable of employing perhaps hundreds of foragers. So here again we see foragers (the scouts) performing a behavior which reduces their own expected foraging success, but boosts that of the colony as a whole.

The perspective of this chapter is to view a honeybee colony as a machine which is designed to extract energy and other resources from the environment. By this view the individual foragers within a colony are simply component parts shaped to contribute in a small way to the larger goal of efficient foraging by a whole colony. Our aim is to disassemble this machine, scrutinize its parts, and, most importantly, discern how they all work together to produce an efficient whole. In the end we will have explored one of the most intriguing general ideas in biology, namely, the concept that a group of cooperating units—be they genes, cells, or whole organisms—can solve the challenges of life far more effectively than can a single unit (Dawkins 1982). We will see that in the case of honeybee foraging, the connection between group operations and high efficiency is primarily a result of sharing information about locations of rich food sources. Thus each individual's decision of where to work is based not on the small amount of information she might gather herself, but rather on the broad, collective knowledge of a colony's entire forager force.

Even though the honeybee's system of social foraging yields efficiency at the level of whole colonies rather than at the level of individual colony members, this does not mean that individual-level selection has been subordinate to colony-level selection. Rather, the two patterns of selection have evidently operated in tandem to shape this foraging process. The critical point here is that it is in the genetic self-interest of each forager to have her colony as a whole be as efficient as possible in foraging. This is so because foragers, like workers in general in honeybee colonies, do not reproduce directly, but do so indirectly via their mother queen (see Chapter 3). One way a worker can help her mother produce reproductives, and so help herself, is by helping the colony forage as efficiently as possible. The larger the pool of resources assembled by the colony, the greater the number of reproductives it can manufacture, and the higher each worker's inclusive fitness. In short, with respect to foraging, what is best for the colony is also best for its members, so workers cooperate closely to help their colony achieve high efficiency.

Colony Economics

Just four resources—nectar, pollen, water, and resin—are needed to support life in a honeybee colony. Nectar and pollen are the bees' foods, their sources of carbohydrates and proteins, respectively. Water is gathered primarily for evaporative cooling of the nest interior on hot days and for diluting honey stores to prepare food for larvae. Resin serves to plug unwanted openings in a nest cavity's walls, reinforce a nest when applied as a thin coating over the wax combs, and protect against microorganisms (see Chapter 9). Before examining how colonies collect these materials, it is valuable to consider the overall economy of a colony, in particular, the total amount of resources consumed annually by a colony and the effort expended in their collection. Such a view emphasizes the massive scale of a colony's foraging operations, a fundamental feature of honeybee ecology.

Because nearly all published observations on nectar and pollen consumption are based on colonies managed for honey production, my discussion of colony economics is based largely on my own recent studies of undisturbed colonies in Connecticut (Seeley and Visscher 1985, Seeley, unpublished data). The populations of these colonies range from a minimum of around 10,000 adult bees in late winter (March) to a maximum of some 40,000 bees in late spring (May-June), just before swarming. These numbers represent a biomass of roughly 1.25 to 5.0 kilograms (7700 bees/kg, Otis 1982a). Because colonies are massive and consume large quantities of food, one can measure their food requirements by monitoring changes in the total weight of a colony, its food reserves, and its nest (hereafter this total will be called "hive weight"). Between late September and late April colonies collect little or no food, and thus this is a time of resource expenditure with colonies steadily consuming their honey and pollen stores and their hive weights constantly falling. As was discussed in Chapter 3, the total mass of food eaten in this winter period is approximately 25 kilograms, of which 1 kilogram is pollen and all the rest is honey.

Estimating a colony's total consumption of honey and pollen for the summer (late April to late September) is harder than for the winter because resources now flow into the colony, counterbalancing losses in colony weight due to food consumption. Fortunately there occur extended periods of cool, rainy weather in the summer during which bees cannot forage. Losses in hive weight at these times indicate the summertime rate of resource utilization. (Actually, they must provide an underestimate. Resources consumed but converted into brood are not represented in the weight losses, and moreover weight losses recorded during bad weather provide no indication of energy expended during foraging.) Drops in hive weight during inclement weather range from 1 to 4 kg/week, averaging about 2.5 kg/week. Given a season of 22 weeks (late April to late September), the total mass of resources consumed over the summer is about 55 kg (2.5 kg/week × 22 weeks). The pollen portion of this total can be estimated by noting that it requires about 130 mg of pollen to produce a bee (Alfonsus 1933, Haydak 1935), and that the average colony population across the summer is about 30,000. As an average bee lives about one month (Sekiguchi and Sakagami 1966, Sakagami and Fukuda 1968), this implies that a colony rears about 150,000 bees each summer over the 5-month season. At about 130 mg of pollen per bee reared, a colony would then require about 20 kg of pollen each summer for brood rearing. Hence the yearly food consumption of unmanaged colonies in Connecticut is approximately 20 kg of pollen and 60 kg of honey (25 kg in winter plus 35 kg in summer). These are, of course, only general estimates; the precise values will vary depending on colony size, climate, and forage abundance. Colonies managed for honey production in Europe and North America have been estimated to rear 150,000 to 200,000 bees annually (Brünnich 1923, Merrill 1924, Nolan 1925) and to consume 15 to 30 kg of pollen (Armbruster 1921, Eckert 1942, Hirschfelder 1951, Louveaux 1958) and 60 to 80 kg of honey (Weipple 1928, Rosov 1944) each year.

The number of trips required to pool these materials and the efficiency of foraging are readily calculated. With respect to pollen, a typical load weighs about 15 mg (Parker 1926, Fukuda et al. 1969) so the collection of 20 kg of pollen requires approximately 1.3 million foraging trips. Given an average flight distance of 4.5 km per trip (Visscher and Seeley 1982), a flight cost of 6.5 J/km (Scholze et al. 1964, Heinrich 1980), and an energy value for pollen of 14,250 J/g (Southwick and Pimentel 1981), the total cost of flying to collect this pollen is about 3.8×10^7 J (1.3×10^6 trips $\times$ 4.5 km/trip $\times$ 6.5 J/km) and the pollen energy value is nearly 2.9×10^8 J. Thus bees achieve an approximately 8 to 1 ratio of energy return in collecting pollen. Parallel calculations for nectar collection, assuming that nectar is on average a 40 percent sugar solution (Park 1949, Southwick et al. 1981) while honey is an 80 percent sugar solution (White 1975), and that a typical nectar load weighs 40 mg (Park 1949, von Frisch 1967, Wells and Giacchino 1968), yield estimates of 3 million foraging trips and a 10 to 1 rate of return on flight energy expended for nectar collection.

In summary, we have seen that food collection by honeybee colonies is an enormous undertaking. Each colony can be thought of as an organism which weighs 1 to 5 kg, rears 150,000 bees and consumes 20 kg of pollen and 60 kg of honey each year. To collect this food, which comes as tiny, widely scattered packets inside flowers, a colony must dispatch its workers on several million foraging trips, with these foragers flying 20 million kilometers overall. With these facts in hand, it should not be surprising to expect that honeybees have been under strong selection for efficiency in the procurement and use of resources.

Recruitment to Food Sources

The ability of honeybees to share information about feeding sites greatly helps their colonies achieve high efficiency in foraging. Whenever a bee discovers a new rich food source, she promptly recruits nestmates to it and so helps ensure that her colony's forager force stays focused on the richest available patches. The principal mechanism of this recruitment communication is the waggle dance, a unique form of behavior in which a bee, deep inside her colony's nest, performs a miniaturized reenactment of her recent journey to a patch of flowers. Bees following these dances learn the distance, direction, and odor of these flowers and can translate this information into a flight to the specified flowers. Thus a waggle dance is a truly symbolic message, one which is separated in space and time from both the actions on which it is based and the behaviors it will guide. Our understanding of this remarkable communication system is due largely to the work of Karl von Frisch and his colleagues, whose investigations now span seven decades. The vast literature on the subject has been summarized in several excellent reviews (von Frisch 1967, Lindauer 1967, Gould 1976, Dyer and Gould 1983). My presentation will emphasize those aspects of the subject which are most relevant to understanding the honeybee's system of social foraging from an ecological perspective.

To examine how bees communicate using waggle dances, let us follow the behavior of a bee upon her return from a rich, new food source. Her find is a large cluster of flowers located a moderate distance from her nest, say 1500 meters, and along a line 40° to the right of the line running between the sun and her nest (Fig. 7.1). Excited by her discovery, she scrambles into her nest's entrance opening and immediately crawls onto one of the vertical combs. Here, amidst a massed throng of her sisters, she performs her recruitment dances. This involves running through a small figure-eight pattern: a straight run followed by a turn to the right to circle back to the starting point, another straight run, followed by a turn and circle to the left, and so on in a regular alternation between right and left turns after straight runs. The straight section of the dance is the most striking and informative part of the signaling bee's performance and is given special emphasis by two behaviors. First, while walking straight ahead the dancer vigorously vibrates ("waggles") her body back and forth laterally, with these sideways deflections greatest at the tip of her abdomen and least at the head. She further emphasizes the straight runs by emitting a buzzing sound during the abdomen-wagging portion of the dance. Produced by the flight muscles and characterized by a frequency (250 Hz) corresponding to wingbeat vibrations, these buzzes evidently represent a ritualized flight-intention behavior. Usually several bees will be tripping along behind a dancing bee, their antennae always extended toward the dancer, trying to maintain contact and acquire information. The followers probably detect the dance sounds with their antennae. The flagellum (distal portion) of a worker bee's antennae has a resonant frequency of 280 Hz, approximately matching the vibration frequency of the waggle-run buzz-

Figure 7.1 The waggle dance of the honeybee. (A) In this example the patch of flowers lies 1500 meters out along a line 40° to the right of the sun as a bee leaves her colony's nest. (B) To advertise this target when inside the nest, the bee runs through a figure-eight pattern, vibrating her body laterally as she passes through the central, straight run. The straight run is oriented on the vertical comb by transposing the angle between the food and the sun to the angle between the straight run and vertical. (C) Distance to the flowers is coded by the duration of the straight run (based on data in Table 14, von Frisch 1967.)

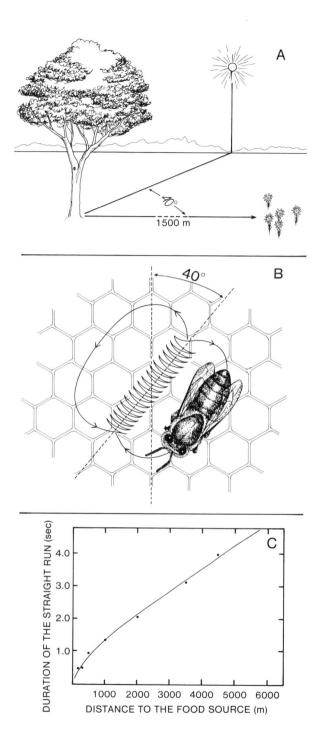

A

B

40°

C

DURATION OF THE STRAIGHT RUN (sec)

4.0

3.0

2.0

1.0

1000 2000 3000 4000 5000 6000
DISTANCE TO THE FOOD SOURCE (m)

1500 m

ing; moreover, the Johnston's organ, a vibration detector at the base of the flagellum, is maximally sensitive to vibrations in the 200 to 350 Hz band.

The direction and duration of straight runs are closely correlated with the direction and distance of the patch of flowers being advertised by the dancing bee. Flowers located directly in line with the sun are represented by waggle runs in an upward direction on the vertical combs, and any angle to the right or left of the sun is coded by a corresponding angle to the right or left of the upward direction. In the example illustrated in Figure 7.1, the flowers lie 40° to the right of the sun, and the straight run of the dance is correspondingly directed at an angle of 40° to the right of vertical. The distance between nest and target appears to be encoded in the duration of the straight runs, since this is the property of the dance which exhibits the tightest correlation with distance to a goal. The farther the target, the longer the straight-run period, with a rate of increase of about 75 milliseconds per 100 meters. It seems likely that dance followers measure the duration of a dancer's straight runs by noting the duration of the 250 Hz buzzing sound produced during each dance circuit.

Besides information about the direction and distance of a rich patch of flowers, a dancing bee also communicates the odor of the flowers at her forage site. In part this scent is carried back in the forager's waxy cuticle, but often a stronger source is the food she brings home—the loads of pollen on her hind legs or the nectar she regurgitates to surrounding bees. Recruits appear to draw upon their knowledge of the food source's odor to help pinpoint its location after using the dance's vectorial information to arrive in the general vicinity. Intensifying the odor at a food source speeds the recruitment process (von Frisch 1967, Johnson and Wenner 1970, Gould 1976). If the recruit target lacks significant odor (as, for example, if it is a water source, a new nest site, or a clump of weakly scented flowers), bees will mark the site with scent from their Nasonov glands (Free 1968, Free and Williams 1970).

There is no question that the waggle dance, a ritualized encapsulation of a journey to a feeding place or other goal, is a truly elegant means of recruitment communication, but just how well does it work? What are its range, precision, and efficiency? Only the first of these topics—the spatial scale of possible recruitment—can be meaningfully answered at present. It is well known that bees will routinely forage several kilometers from their nest, up to a maximum of about 12 kilometers (Eckert 1933, Knaffl 1953, Beutler 1954, Gary, Witherell, and Marston 1972). It is also clear that bees will recruit nestmates to forage patches up to 10 or more kilometers from home (Knaffl 1953, von Frisch 1967, Visscher and Seeley 1982).

The questions of recruitment precision and efficiency are more problematic because most measurements of these matters have been performed with undersized experimental arrays. In nature, honeybees rarely forage within 500 meters of their nest, but in setting up experiments, bee researchers rarely work beyond 300 meters from their hives. Thus the results currently available must be viewed as preliminary.

The general procedure for assessing recruitment precision is to establish a small group of foragers (recruiters) at one feeder, position other feeders in a regular array around the first one, and monitor the arrival of recruits across the array. One refinement of this method involves having the recruiters "misdirect" their recruits to an array of feeders well away from the food source being visited by the recruiting bees (Gould 1975a, 1975b). This eliminates the possibility of the recruiters supplying olfactory or visual signals (though these are appropriate under natural conditions), and thus the observed recruitment pattern reflects waggle-dance recruitment alone. For a target at 150 meters, Gould (1975a) observed an average directional accuracy in recruitment of $\pm$ 11.9° ($\pm$ 31 m). This total error or scatter around the goal appears to reflect nearly equal amounts of scatter in the recruiters' dances ($\pm$ 8.1°) and in the recruits' use of the dance information ($\pm$ 8.7°). (Sum-of-the-squares analysis: $(8.1)^2 + (8.7)^2 = (11.9)^2$; the total scatter and the dance scatter are measured, whereas the recruit error is inferred.) Measurements made with the target at 400 m suggest that as the separation of nest and food source increases, the average directional accuracy improves in terms of angle ($\pm$ 4.1°) but not absolute distance ($\pm$ 29 m). Estimates of recruitment accuracy with respect to distance reveal comparable values in absolute terms (Gould 1975b). For targets 60 to 390 m from the hive, bees showed an average recruit accuracy of $\pm$ 11 percent (roughly $\pm$ 25 m, overall; see Fig. 7.2). Working over somewhat greater distances and with slightly different techniques, von Frisch (1967) likewise observed that most recruits arrive somewhere within a circle of radius 50 m around the advertised feeder (see data analysis in Gould 1976).

Given this level of precision, it seems clear that the actual rate of recruitment to a target will be strongly influenced by such things as its size and its richness in visual and olfactory cues. Thus the efficiency of recruitment, measured in terms of recruits arriving at the goal per bout of dancing by a recruiter, is likely to vary strongly from setting to setting. However, one can generalize by saying that honeybee recruitment is far from an automatic process; bees do not simply follow a few dances and then fly promptly to the target. Repeated measurements of the recruitment rate to feeders positioned only 100 to 400 m from a hive, in the center of an open field, and usually strongly marked with scent, reveal that over a period of several hours only a small minority of the bees which have followed dances for the feeder actually find the feeder (Gould et al. 1970, Esch and Bastian 1970, Mautz 1971). For example, in one series of observations, 1072 bees closely followed dances advertising a food source but only 346 (34 percent) ever arrived there (Mautz 1971). The

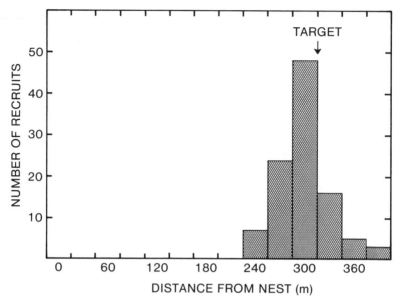

Figure 7.2 Precision of recruitment by the honeybee's waggle dance to a target located 315 meters from the hive. (Modified from Gould 1975b.)

successful recruits required on average 2.4 trips outside to search for the feeder before eventually locating it. In another set of observations, conducted under more natural conditions where recruitment was to patches of flowers 1340 ± 960 m away, even more of these search trips, each one preceded by dance following, were needed: 4.8 ± 3.2 trips (Seeley 1983). It is therefore not surprising that in one study, where recruitment was to a feeder just 150 m away, the recruitment efficiency was only about 0.5 recruits per bout of dancing by a forager (Gould et al. 1970). Nevertheless, the waggle dance recruitment system can generate a rapid buildup of nestmates at a food source through a cascade effect as successful recruits become recruiters. Thus, for example, even with an efficiency as low as one recruit per two bouts of dancing, the number of bees working a forage patch requiring 20 minutes per roundtrip can rise more than tenfold within two hours.

The Information-Center Strategy of Foraging

It is helpful to think of a colony of honeybees as if it were a gigantic amoeba, fixed on a nest site, but able to send pseudopods—i.e., groups of foragers—out across the forest to patches of flowers rich in nectar and pollen. By studying the foraging behavior of this giant amoeba, learning in particular how it distributes its forager groups across the environment, one can begin to understand the overall design of the honeybee's system of social foraging. However, assembling such an overview of a colony's foraging operations presents a formidable challenge. A typical colony sends into the field some ten thousand or more foragers each day, and given that each bee can fly out several kilometers from her nest, these foragers will be scattered far and wide across the countryside. Nevertheless, it has recently proven possible to monitor a whole colony's foraging behavior by tapping into the bees' own "conversations" about their foraging. This involves working with a colony housed in a glass-walled observation hive and reading its recruitment dances to learn where the colony is foraging.

The essential story of colonial foraging behavior is presented in Figures 7.3 through 7.5. These diagrams are extracted from the work of Visscher and Seeley (1982) who monitored a full-size colony living in a deciduous forest in New York State. Figure 7.3 illustrates how the overall pattern of honeybee foraging is one of a rapidly changing mosaic of forage patches. For example, on June 14, the colony whose activities are depicted in Figure 7.3 was collecting nectar primarily from patches in the southwest (2–3 km) and northwest (0.5 km), and gathering pollen principally from two patches in the south (0.5 km) yielding yellow or yellow-grey pollen. Just four days later, on June 18, the colony had almost completely abandoned these patches, and was instead concentrating mostly on a nectar source to the northeast (1–2 km) and sources of brown and yellow-orange pollen to the northeast (2 km) and southeast (4 km), respectively. Thus the forage maps for these two days differ radically, but are linked by the transitions shown in the maps for June 15–17. This striking feature of rapid, mass shifts of foragers among food sources has also been detected in observations conducted at patches of flowers (Darwin 1877, Butler 1945, Weaver 1979). Fourfold rises in forager density from one day to the next have been observed, evidently a product of heavy recruitment to these patches. This point is underscored in Figure 7.4, whose kite diagrams illustrate that the importance of each patch changes daily, so that no two consecutive days exhibit the same distribution of a colony's foraging effort across patches. This figure furthermore emphasizes that colonies work relatively few patches each day (about 10 patches daily) with each patch being worked for only a relatively brief period (about 7 days). Figure 7.5 indicates another important feature of honeybee foraging, namely, that it extends over a vast area around a colony's nest. For the colony studied, the most common forage patch distance was 600 to 800 m; however, the distance distribution is highly skewed so that the mean is more than 2000 m and the circle enclosing 95 percent of the colony's foraging activity has a radius of 6000 meters. Thus a colony's food collection spreads over an area larger than 100 square kilo-

POLLEN CODES

NECTAR ● YELLOW-GREY POLLEN ⊙ YELLOW POLLEN + ORANGE POLLEN △ YELLOW-ORANGE POLLEN X BROWN POLLEN ○

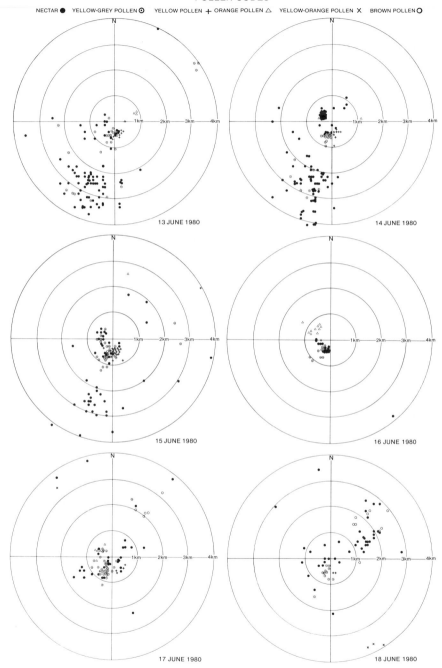

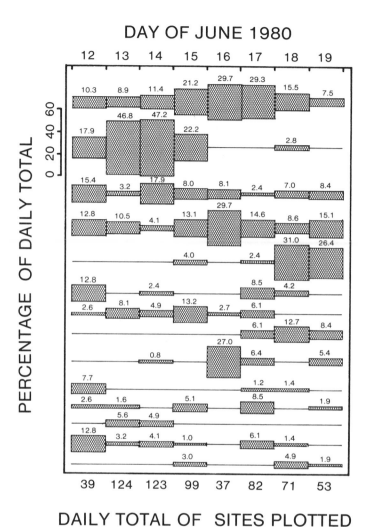

DAY OF JUNE 1980

PERCENTAGE OF DAILY TOTAL

DAILY TOTAL OF SITES PLOTTED

Figure 7.4 Summary of the turnover of food source patches from day to day. Each line represents a patch of forage of a given pollen color. The widths of the lines (and the numbers above them) denote the percentage each patch represents of the total number of sites plotted each day. (Modified from Visscher and Seeley 1982.)

Figure 7.3 Maps of a honeybee colony's daily forage sites, as inferred from reading its recruitment dances. Locations beyond the edge of each map (beyond 4 kilometers) are not shown. (Modified from Visscher and Seeley 1982.)

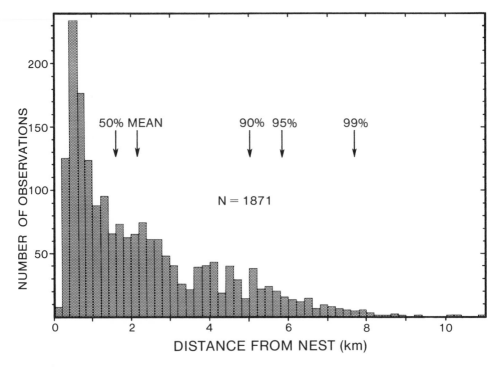

Figure 7.5 Distribution of distances between a honeybee colony's nest and its forage patches. (From Visscher 1982.)

meters. Earlier studies of the foraging range of honeybee colonies suggested that bees rarely work more than 2000 m from their nest (Knaffl 1953, Beutler 1954, Levchenko 1959, Olifir 1969, Gary, Mau, and Mitchell 1972), but these studies employed either unusually small colonies, or were conducted in agricultural areas with unusually rich forage, or both, and so do not reflect natural foraging by honeybees. The foraging behavior of honeybee colonies living in nature therefore is characterized by daily readjustment of the distribution of foragers across patches, relatively few patches used at any one time with each one worked for only a few days, and regular foraging at sources several kilometers from the nest.

These patterns, when combined with the knowledge that honeybees possess recruitment communication, suggest that the basic plan of honeybee foraging involves the colony as an "information center," monitoring a vast area around the nest for food sources, pooling the reconnaissance of the foragers, and somehow using this information to focus a colony's forager force on a few, high-quality patches within its foraging range (von Frisch 1967, Heinrich 1978, Visscher and Seeley 1982).

The heart of this information-center model of honeybee foraging is the idea that colonies organize their foraging so that on any given day a colony's foragers are focused on the few best forage patches that the colony has found. This ability to choose among forage patches has been demonstrated experimentally by offering a colony two sugar-water feeders which differ in concentration of the sugar solution, ease of obtaining the solution, or distance from the colony's nest (Boch 1956, von Frisch 1967, Seeley, unpublished data). Unerringly, colonies concentrate their efforts on the more profitable feeder. For example, a colony's recruitment rates to two feeders, both of which were located 500 meters from the colony's hive and were worked by 30 of the colony's foragers, but whose sugar solutions were 2.25 and 1.50 mol/L, were 79 and 10 recruits in 8 hours, respectively. Experiments with natural food sources also illustrate the sensitivity of colonies to differences in patch quality. When Schaffer et al. (1983) excluded ants from patches of *Agave schottii*, thereby raising the early-morning volume of nectar in the flowers from 2.7 to 16.8 μl/flower, the density of honeybees working these patches rose threefold.

Decision Making by Colonies

The foragers of honeybee colonies closely track the richest available food sources over vast areas around their nests. How do they achieve this remarkable feat? To answer this question we must make a detailed examination of the dense network of interactions among colony members involved in food collection and food storage, for decision making about a colony's food sources is not conducted by some small group of leader bees, but instead is a product of the intricately interwoven behaviors of thousands of individual bees. Specifically, forage patch selection by honeybee colonies is an automatic outcome of the simultaneous operation of the following three basic processes:

(1) workers abandoning patches of relatively low profitability,
(2) workers usually locating new patches by following recruitment dances,
(3) workers recruiting only to patches of relatively high profitability.

The net effect of these behavior patterns working in concert is a steady, gradual migration of a colony's foragers off poor and onto rich food sources.

Assessing the relative profitability of forage patches. Processes 1 and 3 listed above are simply the two extremes in a range of responses to patch quality. When the profitability of a forager's patch is low relative to what many of her nestmates' patches provide, she abandons it, but when her patch is more profitable than most others, she recruits other foragers to her site. Intermediate situations also frequently arise in which foragers continue working their patch but do not advertise it with recruitment dances (von Frisch 1967).

One deep mystery surrounding processes 1 and 3 is how a forager determines the profitability of her patch relative to those of her nestmates' patches. It is clear that honeybees do not do this by visiting other work sites besides their own and drawing comparisons directly. For one thing, major forage sites are usually spaced a kilometer or more apart (Fig. 7.3), and thus simply finding other sites, whether by random searching or following dances, would require much time. For another, when foragers have been individually labeled and then monitored for several days while working a feeder (von Frisch 1967) or flowers (Ribbands 1949), it has been observed that most bees work steadily at their food source without taking ''breaks'' to inspect alternative sites. It is also clear that honeybees do not judge relative patch quality through reference to a fixed, internal scale of quality. This fact is demonstrated by dramatic changes in the definition of a top-quality food source, that is, one which elicits recruitment. For example, when many plants are in bloom and so forage is abundant, a bee feeder must contain a 1.5 to 2.0 mol/L sucrose solution for it to rank highly enough to receive recruits, yet at times of sparse forage the same feeder loaded with just a 0.125 to 0.25 mol/L solution will attract numerous recruits (Fig. 7.6).

One promising approach to solving the mystery of how honeybees assess the relative profitability of their forage patches is to employ certain concepts from optimal foraging theory (reviewed by Pyke et al. 1977, Krebs 1978, 1980, and various authors in Kamil and Sargent 1981). Within this body of ideas, the models of central-place foraging theory (Orians and Pearson 1979, Schoener 1979, Orians 1980) are especially relevant to honeybee foraging. I will apply these ideas to just one foraging situation, the one which is most amenable to an energetic analysis, namely, the case of bees foraging just for nectar. This is not a rare phenomenon for honeybees. Of more than 13,000 bees followed by Parker (1926) as they foraged on 31 plant species, 58 percent were nectar specialists, 25 percent gathered only pollen, and the remaining 17 percent collected both nectar and pollen. Free (1960) likewise found that among bees pollinating fruit trees, only 16 percent collected both types of forage simultaneously.

It seems reasonable to assume that what a honeybee foraging for nectar would like to maximize is her net rate of energy intake. Pyke (1981) discusses the validity of this assumption for nectivores in general, but it is probably particularly true for honeybee foragers because they are sterile members of insect colonies which possess recruitment communication and have large energy budgets. These special circumstances imply that when a honeybee worker is foraging she need not simultaneously be conducting other activities which for solitary insects might conflict with efficient foraging, such as look-

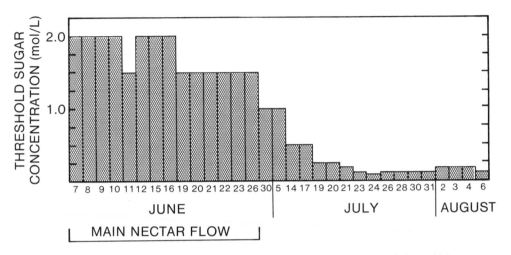

Figure 7.6 Seasonal changes in the threshold concentration of a sugar solution which elicits recruitment dances. These changes in threshold demonstrate that honeybees do not judge nectar sources through reference to a fixed scale of quality. (Modified from Lindauer 1948.)

ing for mates or oviposition sites. They also imply that nectar foragers can specialize very strictly on nectar collection since other bees will cover the colony's pollen needs. Furthermore, foragers engaged in working a profitable patch need not devote time to sampling alternative sources to monitor possible changes in foraging opportunity. Such sampling is performed by other bees, the colony's scouts. And finally, the social context of honeybee foraging means that workers will nearly always benefit from collecting as much energy as possible. Honeybee colonies not only consume vast amounts of energy, but also their capacity for growth, reproduction, and especially winter survival is frequently limited by their energy supply (Seeley 1978, Seeley and Visscher 1985).

To understand the various factors influencing a nectar forager's rate of energy intake, it is helpful to think of a forage patch in terms of three basic parameters: (1) the time needed to fly to and from the patch (travel time, T_T), (2) the time needed to load up with nectar when in a patch (patch time, T_P), and (3) the energy gained per trip to the patch. The relationships among these three variables and overall patch profitability can be represented graphically, as shown in Figure 7.7. The slope of line $\overline{AB}$ describes the rate of energy intake realized by a forager in working a patch. Note that although this slope represents the gross rate of energy intake per trip, not the net rate, the difference between the two—the cost of traveling to and from the patch—is generally small. Figure 7.7 is drawn based on realistic values for the relevant

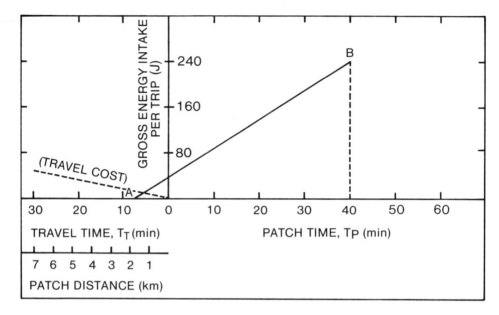

Figure 7.7 Graphical representation of the calculation of a forager's rate of energy intake (indicated by the slope of line $\overline{AB}$) from a forage patch which is 1.7 kilometers from the bee's nest, offers a 40 percent sugar solution, and requires 40 minutes at the patch for the forager to collect a full 40 milligram load of the sugar solution. The dashed, sloping line on the left side denotes the cost of a roundtrip flight between nest and forage patch for various distances, assuming foragers fly 7.7 m/sec (von Frisch and Lindauer 1955, Boch 1956), weigh 100 mg (Otis 1982a), and have a metabolic rate while flying of 503 W/kg (Heinrich 1980). This line indicates that for forage patches up to about 5 kilometers from the nest, the flight cost of a forage trip is only a small fraction of the trip's gross profit.

parameters (sugar concentration of nectar: 40 percent [Park 1928, 1949, South-wick et al. 1981]; nectar load: 40 mg [Park 1928, 1949, Fukuda et al. 1969]; and travel cost: 6.5 J/km [Scholze et al. 1964, Heinrich 1980]), and it illustrates that travel costs typically total but a few percent of the gross energy intake per trip. For example, given the parameter values stated above, then for an average patch located 1.7 km from a bee's nest, the travel costs total just 9 percent of the gross energy gained per trip.

Figure 7.7 indicates that what a nectar forager must know to estimate the relative profitability of her patch is the relative energy gained and relative time spent ($T_T + T_P$) in making a trip to this patch. How might a forager determine these values? The relative (gross) energy gained per trip is a function of two factors—the volume of nectar imbibed and the sugar concentration of the nectar—only one of which, however, may be a variable. Because worker honeybees show little size variation (Kerr and Hebling 1964, Waddington 1981), a full nectar load is similar for all foragers. But do nectar foragers consistently fill up to the same degree? The answer remains unclear. Several studies (Núñez 1966, 1970, 1982, Wells and Giacchino 1968) indicate that the volume of a nectar load is unaffected by the nectar's sweetness, but that it increases significantly as patch time decreases and travel time increases. However, these are the feeding patterns of bees foraging at artificial food sources, and they may bear little resemblance to the behaviors found at flowers. Fukuda et al. (1969) weighed the nectar loads that bees returned to their nests and found wide variation in load size, but their samples undoubtedly included many bees not fully engaged in nectar collection, such as scout bees (see below), novice foragers, or experienced foragers in the process of changing work sites. What we really need are measurements of nectar-load volumes for bees actively working patches of flowers.

Assuming that the volumes of nectar loads are basically uniform, then the principal variable determining the energy gained per forage trip is the nectar's sugar concentration. Nectars gathered by honeybees range between about 10 and 70 percent sugar (Oertel 1944, Beutler 1953, Percival 1965, Southwick et al. 1981). How can a forager determine how her nectar's concentration compares with those of her nestmates' nectars? Studies by Lindauer (1948, 1954) revealed a communication system involving feedback between foragers and hive bees which transmits precisely this information to foragers. When a nectar forager returns to her nest, she delivers her load to a receiver bee, one of a group of workers specializing in the reception and storage of nectar. Because each receiver unloads foragers working various forage patches (Nowogrodzki 1981), they know the range of sweetness for incoming nectars, can rank particular nectar loads on a scale of sugariness, and adjust their behavior while unloading a forager accordingly. Foragers with relatively concentrated nectar evoke a strong reception and are speedily unloaded, whereas ones with relatively dilute nectar arouse little interest and often must search hard to find a receiver willing to accept her second-rate forage. Foragers clearly register this information about relative sweetness, for Lindauer observed that a bee's tendency to dance after unloading correlates closely with her delivery time. If this is less than 40 seconds, then bees nearly always dance for their patch, but as this time increases the dance tendency declines sharply so that dances rarely follow deliveries requiring 100 or more seconds. Boch (1956) uncovered a further feature of this elegant system of communication, namely, that the receivers' responsiveness to differences in sugar concentration reflects the abundance of forage. The heavier the flow of nectar into the nest, the choosier are the receivers in deciding whom to unload speedily. This neatly ensures that colonies focus on top-quality sources when they enjoy a wide selection, but also that they do not bypass poorer sources when few forage patches are available. This communication system is remarkably effective. As is shown in Figure 7.8, food sources whose sugar solutions differ by as little as 0.125 mol/L elicit significantly different rates of recruitment.

The mechanisms just discussed evidently enable nectar foragers to estimate the relative energy gained per trip to a patch, but what about the relative time per trip? Both values are needed to estimate the relative rate of energy intake. Bees clearly take trip time into account when deciding about recruitment to a patch. Thus, for example, when bees from one colony were trained to collect from two feeders, both containing 1.5 mol/L sugar solution, but one 100 m (trip time about 85 sec) and the other 525 m (trip time about 190 sec) from the bees' hive, fully 59 percent of the near-feeder bees but only 34 percent of the far-feeder foragers advertised their feeder with dances (Boch 1956). Also, Waddington (1982) found that bees foraging from artificial flowers in a laboratory arena dance less vigorously when the flowers are spaced farther apart and thus the time (and energy) spent in gathering a nectar load is

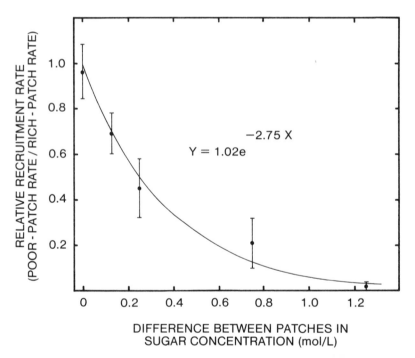

Figure 7.8 Sensitivity of a honeybee colony to differences in sugar concentration between forage patches. A negative exponential describes a colony's response to linear increases in the difference in sugar concentration between patches. (From Seeley, unpublished data.)

increased. It is hard to see how foragers can learn the durations of nestmates' foraging trips as a basis for estimating the relative length of their own trips. Perhaps they instead rely upon an innate scale of trip times through which a bee recognizes that foraging trips lasting 20, 40, and 80 minutes, for example, are relatively short, medium, and long, respectively (Park 1928, 1949, Singh 1950, Ribbands 1951).

Before concluding this discussion of how bees assess the relative profitability of forage patches, I must point out that the model of patch profitability presented in Figure 7.7 is certainly oversimplified. For example, besides trip time, patch time, and energy per trip, bees also enter a patch's riskiness into their calculations of patch quality. Thus when a thunderstorm approached one of Boch's (1956) experimental arrays, he observed that bees working a feeder 6000 m from their hive had completely abandoned it by 18:09 hours, but that bees foraging from a feeder just 100 m from the hive continued working until 19:15 hours, when the rains began to fall. The model also overlooks the fact that the net energetic value of a sugar solution is not directly proportional to its concentration. Nectar must be concentrated to form honey and this activity costs bees less the higher the nectar's initial sugar concentration. But despite such shortcomings, this central-place foraging model provides a useful preliminary framework for organizing information on how bees judge patch quality.

Division of labor between scouts and recruits. For honeybee colonies to keep their foragers trained on rich food sources, their workers must not only selectively recruit to the best sources, they must also rely primarily upon the recruitment process when they need to find new forage patches. However, not all of a colony's foragers can be recruits, the exploiters of known sources; some must be scouts, the explorers for unknown patches of flowers.

Oettingen-Spielberg (1949) made a preliminary measurement of the percent scouts among novice foragers by adding marked, newly emerged bees to a hive kept in a flight room and observing what fraction of these bees began foraging by arriving at a feeding station, to which there was recruitment, or to a bouquet of flowers, to which no recruitment was allowed. Of 1062 bees, only 53 (5 percent) found the flowers and so presumably were scouts. Lindauer (1952) repeated this measurement using a far superior experimental technique. He observed which individuals within a large, single-age cohort of bees inhabiting an observation hive started foraging without following any dances. He observed 13 percent scouts (data analyzed in Seeley 1983). I have recently extended these measurments to estimate the percent scouts among experienced foragers (Seeley 1983). Out of 78 bees, 18 (23 percent) were scouts. Clearly, the large majority of honeybee foragers, both novice and experienced, find food sources by following recruitment dances.

The figure of 23 percent scouts among experienced foragers is not a fixed percentage, but is actually adjusted in relation to forage abundance. During a dearth, the proportion of scouts rose to approximately 35 percent, but just two weeks later, when a profusion of catnip (*Nepeta cataria*) flowers burst into bloom, this figure plummeted to 5 percent. Such flexibility seems highly adaptive. When forage is poor and many foragers are unemployed, strong investment in finding additional food sources seems appropriate. Conversely, when forage is plentiful, rich sources are easily found, so colonies need not invest heavily in searching, but instead should devote most of their foragers to harvesting the plentiful food.

Energetic consequences of social patch choice. What are the benefits and costs of the honeybees' information-center strategy of foraging? Consider first the spatial effects of their foraging system. The average rate of food collection per forager is probably raised by this social organization because each colony, foraging as an integrated whole, can sample widely through the environment and can focus foragers on the best patches. For solitary insects or even social insects without recruitment communication, such as bumblebees (Heinrich 1979a), each individual's knowledge about food sources is limited to what it can collect personally and the area it surveys must be but a tiny fraction of the 100 or more square kilometers monitored by an entire honeybee colony. Evidence supporting these ideas has been obtained by comparing the foraging patterns of honeybee recruits, which use the information exchange, and scouts, which do not (Seeley 1983). As expected, recruits seem to find better food sources than do scouts. Fully 95 percent of recruits, but only 51 percent of scouts, which have recently located a new food source find it worthwhile to make a second visit to their find. However, locating superior patches through social searching evidently involves higher search costs. Recruits required 138 ± 76 min outside their nest to find a new food source in comparison with only 85 ± 58 min needed by scouts. Recruits search longer probably because they are trying to locate one particular patch of flowers, the one specified by a nestmate's dances, and, as we have seen earlier, locating forage patches by following dances can be time-consuming. Scouts, in contrast, do not seek any one special patch, but whatever patch they should encounter by chance. Finding flowers this way evidently requires relatively little search effort. It seems likely, though, that recruits come out ahead of scouts in overall foraging effectiveness since the difference in their search costs is only on the order of 120 joules, less than the profit of one average foraging trip, an amount which is probably quickly counterbalanced by recruits working far superior food sources.

The honeybees' social organization for foraging may enhance their efficiency in food collection through temporal as well as spatial effects. Insofar as rich food sources are available only briefly, either because the flowers in a patch bloom just for a short time or because patches are quickly drained by whoever first discovers them, a colony will benefit from being able rapidly to direct foragers to work sites and from having individuals (scouts) specializing in monitoring the countryside for new food sources. Overall, though, I feel the honeybee's system of social foraging bears the stamp of design for effectiveness in locating and tracking top-quality food sources more than design for speedy exploitation of ephemeral sources. Just one sign of this is the way scout bees generally perform several visits to a newly discovered food source before starting to recruit nestmates there (Lindauer 1948). Apparently there is more of a premium on sending recruits to the right locations, if somewhat slowly, than on rapidly getting them onto sites, only some of which may prove truly worthwhile.

Colonial foraging decisions besides patch choice. In collecting its food, a honeybee colony faces the following nested hierarchy of foraging decisions. First, what fraction of a colony's total work force should be devoted to food collection? Second, given a forager force of a certain size, how should a colony partition it between scouts and recruits? Third, given a certain number of recruits, how should they be divided between the tasks of pollen and nectar collection? And fourth, given a certain number of recruits for either pollen or nectar collection, how should they be distributed among the known patches yielding either pollen or nectar?

So far we have examined only the last of these decision-making processes in detail, but it appears that colonies also make decisions at the other three levels of the foraging process. The fraction of a colony's work force which is devoted to foraging reflects forage abundance. When forage is plentiful, workers start foraging at an earlier age than when little forage is available (Lindauer 1952, Sekiguchi and Sakagami 1966), thereby creating a larger proportion of foragers in a colony during good times than bad times. Similarly, as we have seen above, the proportion of scout bees in a colony's forager force reflects forage availability. The key to understanding how colonies regulate their scouting effort in relation to forage abundance may be found in the variation among individual foragers in the tendency to perform as a scout or recruit. The idea here is that some bees always scout to find a new forage patch, others always follow recruitment dances, and still others employ either technique. What determines whether or not these intermediate, versatile foragers follow recruitment dances may be simply the ease of finding dancing bees to follow. If so, then as the availability of forage decreases, and the number of bees advertising underexploited forage patches dwindles, a colony will automatically allocate more bees to scouting (Seeley 1985). In turn, this should help boost the colony's probability of discovering additional food sources. Supporting this hypothesis are the numerous observations of consistent differences among individual foragers in such things as rate of foraging, sensitivity to disturbance at a feeder, tendency to scout old sources, and inclination to seek an alternative food source when an old one starts deteriorating (von Frisch 1923, Opfinger 1949, Schmid 1964, Seeley 1983).

Labor allocation between pollen and nectar collection also appears to be regulated in accordance with a colony's needs. When Lindauer (1952) equipped one colony's hive with a pollen trap (an apparatus which scrapes the pollen loads from a portion of foragers as they enter their hive), but left two other colonies undisturbed as controls, he observed over the following seven days that the percent foragers returning with pollen rose from 15 to about 80 percent for the experimental colony, whereas it declined from 10 to 1 percent in the control colonies (Fig. 7.9). Also, when Free (1967b) doubled the amount of brood in colonies, he found that their rate of pollen collection more than doubled, though over the same time interval colonies which did not receive additional brood reduced their pollen collection by about one-fifth. Free (1967b, 1970) speculates that a colony's nurse bees signal a shortage of pollen to the colony's foragers by preparing empty cells to receive pollen. The pollen foragers, in turn, may sense the level of need

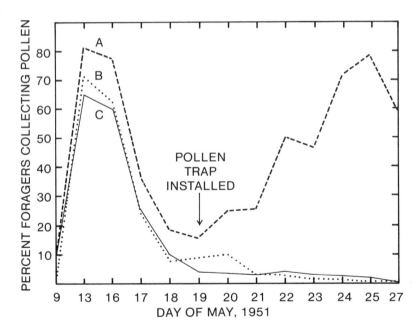

Figure 7.9 The influence of a pollen shortage on the percent pollen foragers in a colony. In colony A, a pollen shortage was induced by installing a pollen trap on 19 May. Colonies B and C were left undisturbed as controls. (From Lindauer 1952.)

for pollen by the ease with which they find cells prepared to receive their pollen loads. When such cells are quickly found, pollen foragers may be stimulated to continue collecting pollen and perhaps even recruit nestmates to this task by dancing for their pollen sources.

Decisions by Individuals in a Flower Patch

Once a forager succeeds in locating a rich patch of flowers, guided in her search by the recruitment and other social processes already discussed, she must make several decisions on her own in order to exploit her new food source efficiently. If several types of flowers are in bloom at the patch, she must decide which ones she will work. Having made her choice, she must then plot her moves among individual blossoms.

It is clear that honeybee foragers make firm decisions about which type of flower to visit; almost invariably, each forager specializes totally on the flowers of one species. This is evident from observations of bees in flower gardens steadily picking their way from blossom to blossom of a single species (Darwin 1877, Ribbands 1949), but is even more vividly illustrated by analyses of the pollen loads of foragers. The overwhelming majority—95 to 99 percent—contain just one type of pollen (Betts 1920, 1935, Maurizio 1953, Free 1963). This tendency to specialize is so strong that when offered an experimental array of artificial flowers containing several flower types, honeybees will lock in quickly on one type of flower, thus often overlooking a second, even richer "species" of flower (Wells and Wells 1983). In nature, such specialization enhances foraging efficiency by helping bees gain high proficiency in working flowers, learning such things as how to recognize fresh flowers, where exactly to alight on flowers, and how to quickly insert a tongue to suck up nectar or how to scrabble over a flower's anthers to gather pollen. For example, when foragers work hairy vetch (*Vicia villosa*), a plant with zygomorphic flowers whose nectar is concealed from bees, on their first visits they simply land on various flowers and attempt to insert their tongue at random into each flower. Eventually they succeed in reaching the hidden nectar, whereupon each bee's behavior suddenly becomes tightly patterned, with each forager methodically using whichever approach happened to prove successful. Some bees learn to insert their tongues between the standard and keel petals at the base of flowers, while others reach the nectar through the mouth of a vetch flower, using their head as a wedge to spread the standard and keel petals, pressing down on the keel with their legs, and inserting their tongue straight down the corolla tube. Only about 3 percent of the bees employ both approaches (Weaver 1956). The ability of foragers to discrimate different flower types and to focus their labors on one species reflects their skill in learning the scent, color, and shape of their chosen blossoms (reviewed by Lindauer 1970, Menzel et al. 1974). After just one rewarded visit to a flower, foragers will select flowers of the same scent with over 90 percent accuracy, and achieve virtually 100 percent accuracy after just three more successful visits. Flower discrimination based on color or form develops several times more slowly, requiring 5 to 20 rewards before 80 percent accuracy is achieved, but because bees typically collect forage from several dozen flowers per foraging trip, it is clear that within one trip a forager assembles a sharp, multi-dimensional memory of her particular food plant's flowers.

How is this plant species chosen? Often the flower patch to which a bee is recruited contains just one type of flower in abundance, so frequently the flower type is specified by patch location. Other times, however, the blossoms of several species intermingle and a choice is necessary. Laboratory experiments in which bees are offered two varieties of artificial flowers, with the two types differing in their rewards, have demonstrated that honeybees can sample across an array of flower types and focus with fair accuracy on the most profitable form (Waddington and Holden 1979). Such studies, however, evidently examine a relatively rare method of forage selection by bees, the one employed by scout bees. As we have seen, the vast majority of honeybee foragers are recruited to their work sites. These recruits seem to seek flowers in their forage patch whose scent matches the floral odor each bee sensed while following recruitment dances. The precision of this scent matching is remarkably high. Koltermann (1969) found that when recruits who had followed a dancer bearing geraniol scent reached the feeder and were given a choice between geraniol- and fennel-scented dishes of food, they selected the dancer-borne scent with 99 percent accuracy. Thus for honeybees, even a forager's choice of which flower species to work is only properly understood when viewed as part of a social system of foraging.

Having located a patch of flowers, and selected a flower species within the patch, a forager must finally choose a pattern of movement among blossoms of this species. As a rule, the richest flowers will not be uniformly distributed throughout the patch, either because of genetic variation among plants, differences in soil conditions, or, perhaps primarily, patchy use of the flowers by prior foragers (Pleasants and Zimmerman 1979, Zimmerman 1981). Thus to work the flowers efficiently, a forager should not move about haphazardly. If outstandingly rich flowers are rare, and they renew their reward sufficiently rapidly, then a forager might memorize these blossoms' positions and systematically harvest their forage by "trap-lining," that is, by visiting a series of flowers in a fixed sequence. Although this has been observed with honeybees, such as those observed by Ribbands (1949) which were collecting pollen from Shirley poppies (*Papaver rhoeas*), it is rare. Far more commonly, a forager faces a vast number of indistinguishable blossoms, hundreds of which must be visited to assemble a full load. Foragers evidently cope with this situation by adjusting the directionality and distances of inter-flower moves so that they linger in areas of rich flowers and pass through poorer regions (Heinrich 1983). When bees foraged among artifical flowers positioned in a

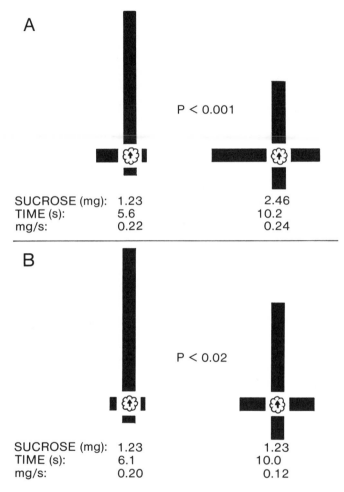

Figure 7.10 Frequency distributions of honeybees' movement directions after visiting artificial flowers with the properties of sucrose/flower and loading time/flower shown. Arrows indicate the direction forward. The comparison in part A illustrates that flight directionality decreases as bees visit flowers with more sugar and a longer handling time per flower. The comparison in part B indicates that the differences in flight directionality trace mainly to differences in handling time. (Modified from Schmid-Hempel 1984.)

checkerboard array of rewarding and nonrewarding clusters of flowers, they tended to reduce the change in direction between successive moves (thus tended to move straight ahead) after hitting a string of nonrewarding flowers. This response, together with increases in the lengths of inter-flower moves following frequent encounters with nonrewarding flowers, helped these bees stay clear of the unprofitable flowers (Waddington 1980). Various mechanisms can account for decreased directionality with increasing flower quality, including making more turns when on richer flowers, thus decoupling a bee's landing and takeoff orientations; spending more time on better flowers and so forgetting the prior travel direction; or simply running a motor program which specifies less directionality when flower quality increases (Waddington and Heinrich 1981). Recent studies by Schmid-Hempel (1984) which controlled flower quality while increasing handling time support the second hypothesis; bees appear to forget gradually their arrival bearings the longer they work on a flower (Fig. 7.10).

The ultimate product of these modulations of flight directionality and distance is the honeybee's characteristic movement pattern while foraging. So long as the rewards remain high, individuals generally work a small, 10 to 40 m² forage area, visiting nearest-neighbor flowers; but when conditions deteriorate, bees become "restless," jumping long distances between successive flowers and so sampling new portions of their flower patch (Butler et al. 1943, Ribbands 1949, Singh 1950, Weaver 1957).

8 Temperature Control

Sociality and the Origins of Nest Thermoregulation

Precise control of nest temperature can be regarded as one of the major innovations in honeybee biology made possible by the evolution of their societies. From late winter to early autumn, the annual period of brood rearing by honeybees, the temperature in the central, nursery region of each colony's nest is stabilized between 33° and 36°C, averaging about 34.5°C and usually varying by less than 1°C across a day (Hess 1926, Himmer 1927, Dunham 1929). The high thermal stability of the broodnest area is perhaps most strikingly illustrated by the extreme temperature sensitivity of honeybee brood, which has adapted to this stable environment. When Himmer (1927) reared capped brood (pupae and late-stage larvae) in an incubator, he found that few bees emerged at below 28°C or above 37°C, that bees reared at 28–30°C reached adulthood but frequently possessed shriveled wings and malformed mouthparts, and that normal metamorphosis occurred only at the temperatures of 32–35°C. When colonies are without brood, from late autumn to mid winter, the temperature of the bees clustered inside their nest drops somewhat, but remains well above freezing. The core temperature of a broodless winter cluster never falls below 18°C while that of the cluster's mantle stays above about 10°C, even in the face of ambient temperatures of −30°C or colder (Gates 1914, Hess 1926, Simpson 1950, Owens 1971). The mechanisms underlying this impressive thermoregulatory ability comprise a set of tightly integrated behaviors and physiological devices whereby colonies regulate both the production of heat through metabolism and the loss of heat to the environment. Before examining these control mechanisms, we will review the evolutionary origins of nest temperature regulation and its ecological significance.

Honeybees need both to heat and to cool the nest, but in temperate climates at least, generating and preserving heat is the more pressing thermal demand. The ability of honeybee colonies to maintain a warm microclimate inside their nests is a direct extension of the adaptations of individual bees for flight. Being insects, bees fly by flapping their wings—the most energetically demanding mode of animal locomotion—and the flight muscles of insects are among the most metabolically active of tissues (Bartholomew 1981). A flying honeybee expends energy at a rate of about 500 W/kg (Jongbloed and Wiersma 1934, Heinrich 1980). In comparison, the maximum power output of an Olympic rowing crew is only about 20 W/kg (Neville 1965). Not only do flying bees consume prodigious amounts of energy, they also generate a great deal of heat. The efficiency of an insect's flight apparatus in converting metabolic fuel to mechanical power is only about 10 to 20 percent, and thus more than 80 percent of the energy expended in flight appears as heat in the muscles (Kammer and Heinrich 1978). In relatively large insects, such as honeybees, the rate of heat loss is sufficiently low so that during sustained flight an individual's body temperature rises well above that of the surrounding air. The size and metabolic rate of flying honeybees, for example, combine to make a bee's thoracic temperature (T_{Th}) typically 10–15°C above ambient temperature (T_A) at T_A of 15–25°C (Esch 1960, Heinrich 1979b).

Thus in honeybees an elevated body temperature is an obligatory consequence of flight, but more importantly for understanding the origins of their colonial thermoregulation, it has become essential for flight. Worker honeybees must maintain T_{Th} above about 27°C in order to fly (Esch 1976, Heinrich 1979b); flight muscles cooler than this simply cannot generate the minimum wingbeat frequency and power output per stroke needed for takeoff and flight (Josephson 1981). This high minimum T_{Th} for flight evidently reflects two design constraints on flight-muscle enzymes: they must withstand the high thoracic temperatures produced by flight, but when built with sufficient intramolecular bonds to resist denaturation at high temperatures, they are too rigid to operate efficiently at low temperatures (Heinrich 1977, 1981b). Thus while evolving biochemical machinery adapted to high temperatures, honeybees apparently also developed the ability to conduct preflight warm-up, without which they would remain grounded once cooled below about 27°C. The motor pattern of bees warming up their flight muscles consists of simultaneous activation of the wing-elevator and wing-depressor muscles. These muscles therefore contract isometrically, producing much heat but few or no wing vibrations (Esch 1960, 1964).

This preflight warm-up behavior evidently set the stage for the evolution of nest warming, since the mechanisms for warming flight muscles and heating nests are identical. Recordings of thoracic temperatures of bees preparing to leave on foraging flights and of bees heating their colony's broodnest reveal a similar pattern of a 2–3° C/min rise in temperature (Esch 1960) (Fig. 8.1) and in both situations a bee's wings remain motionless, folded over the abdomen. A tight correlation between action potentials recorded from the thorax and contractions of thoracic muscles suggests that the power output (and thus the heat production) of a worker bee's flight muscles is under direct neural control, perhaps through regulation of the rate of dorsoventral muscle

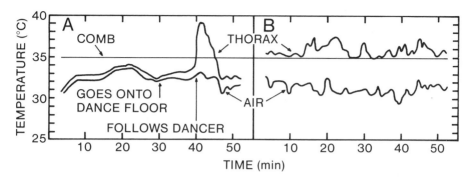

Figure 8.1 Thorax, comb, and air temperature recordings for a bee following recruitment dances and preparing to launch into flight (A), and for a bee heating her colony's broodnest (B). The ability to perform preflight warm-up preadapted honeybees for evolving the ability to heat their nests. (Modified from Esch 1960.)

contractions (Bastian and Esch 1970). Whatever the precise pathway of control, it appears that bees have evolved the ability to generate heat as needed, for preflight warm-up, and that this ability provides the foundation for regulated heating of honeybee nests.

The full development of the honeybees' thermoregulatory skills, however, came only with the evolution of their societies. One of the links between group living and improved nest temperature control is simply the additive capacity for heat production when individuals group together. A colony of 20,000 bees presumably can generate heat some 20,000 times more powerfully than can a single bee. An equally important connection between sociality and thermal homeostasis is the ability of colonies to perform several tasks simultaneously, rather than serially (Oster and Wilson 1978). Thus, whereas a solitary insect can work on nest thermoregulation only part of the time, when, for example, not nest building or foraging, a colony of bees can perform all important tasks concurrently and thus can always have individuals engaged in nest temperature control. Perhaps the greatest advantage in thermoregulation enjoyed by groups over individuals, though, derives from their ability to crowd together into a cluster, through which individual bees greatly reduce the surface area exposed to the general environment and so dramatically reduce their heat loss. The surface area of an isolated bee is about 2 cm² (bee modeled as a cylinder 14 mm long and 0.4 mm in diameter), but the surface area of 15,000 bees, when contracted into a tight winter cluster 18 cm in diameter, is only about 1000 cm², or merely 0.07 cm²/bee, a diminution of effective surface area by a factor of about thirty.

Benefits of Nest Temperature Control

The evolutionary histories of nest microclimate control and sociality are intricately interwoven for honeybees. On the one hand, as we have seen, group living enhances nest thermoregulation. On the other hand, once honeybee colonies evolved to a size of several tens of thousands of individuals, nest temperature control evidently became essential for social life. With thousands of metabolically active adult and immature bees crowded together inside an enclosed nest, a honeybee colony faces the threat of disastrous overheating when ambient temperatures rise above about 30°C. Sustained broodnest temperatures over 37°C disrupt larval metamorphosis (Himmer 1927). Moreover, should the temperature inside a nest rise above about 40°C, the beeswax combs heavily laden with honey will soften and collapse (Chadwick 1931, Seeley, unpublished data). Also the adult bees in a colony can survive only a few hours at temperatures of 45–50°C (Allen 1959a, Free and Spencer-Booth 1962, Heinrich 1980), which is just 10–15°C above their optimum temperature for full activity (35°C). In contrast, honeybees can survive several days at 15°C (Free and Spencer-Booth 1960). Thus honeybees, like most organisms, possess a smaller range of heat tolerance above their optimum than below it. The relatively low tolerance of heat by honeybees may reflect their not evolving enzymes more stable than are normally necessary, since an enzyme which is stable at temperatures far above an organism's normal thermal range is probably too rigid to function efficiently at standard temperatures (Brock 1967).

The adaptive significance of preventing nest overheating is clear, but what selective forces favored the evolution of nest warming? The initial benefit was probably a reduction in maturation time for brood. Because honeybees now develop normally only within a narrow range of temperatures, it is impossible to demonstrate experimentally the effect on development time of a large (10°C) elevation in brood temperature; nevertheless, significant acceleration of development is observed when brood is warmed just slightly. Milum (1930) observed that brood located on the perimeter of a colony's broodnest, where the temperature averaged about 31.5°C, required 22–24 days between egg laying and adult emergence, whereas brood in the nest center was about 3°C warmer and required only 20–22 days to complete development. Similarly, Jay (1959) found that bees reared in an incubator at 32° or 35°C remained as larvae for 10–12 or 7–9 days, respectively. Speedy brood development fosters rapid colony growth and thus quick recovery of full colony size whenever a colony's population has declined, such as after winter, following swarming, or in the wake of heavy mortality from predation or disease. Colonies probably place a premium on rapid growth to full size for many reasons, including earlier swarming, superior defense against predators, and greater food collection.

Once a rudimentary level of nest warming was achieved, the stage was set for several other selective forces to help further propel the evolution of a steady, elevated nest temperature. One of these agents of natural selection was probably disease. Kluger (1979) reviews the evidence that fever—the maintenance of an elevated body temperature—is effective in disease control for fish, reptiles, and amphibians, as well as birds and mammals. The effects of temperature on disease resistance in honeybees, or insects in general, have received little attention (Tanada 1967), but are probably great. It has long been known (Maurizio 1934) that chalkbrood, a disease of honeybee larvae caused by the fungus *Ascosphaera apis*, is blocked if brood is kept warm (35°C). Chilling larvae to 30°C for just a few hours allows fungal spores ingested with food to germinate in a larva's gut and begin to grow mycelia (Bailey 1967). Honeybees suffer from at least fifteen viral and two bacterial diseases (Bailey 1981), but the effect of high broodnest temperature on the honeybee's vulnerability to viral and bacterial infections remains unknown. Studies with other insects, however, suggest the effects are strong. As reviewed by Watanabe and Tanada (1972), various insect viruses fail to cause lethal infection when their hosts are reared at temperatures no higher than those of the honeybee's broodnest. These include a granulosis virus in *Pieris rapae* reared at 36°C, a nuclear polyhedrosis virus in *Diprion hercyniae* at 29°C, and a cytoplasmic–polyhedrosis virus in *Colias eurytheme* at 35°C. Bacterial infections of insects can trigger immune responses (Ratner and Vinson 1983), some of which are temperature sensitive (Chain and Anderson 1983), so the honeybee's elevated nest temperature may also prove beneficial in resisting bacterial diseases.

Still another selective force shaping the honeybee's nest-warming abilities is simply deadly cold. Through its advanced techniques of colonial thermoregulation, the honeybee has greatly expanded its thermal niche, living today in environments where it would otherwise perish each winter. As was discussed in Chapter 4, the honeybee is originally a tropical insect whose range has expanded into cold-temperate regions through various adaptations, especially the ability to maintain a warm cluster throughout long, freezing winters. Because certain tropical races of *Apis mellifera* appear to lack this ability (Taylor 1977), some of the mechanisms of temperature control inside winter clusters (discussed below) are evidently relatively recent adaptations, shaped during the honeybees' penetration into colder regions of the world.

Heating the Nest

The primary problem in thermoregulation faced by honeybee colonies is one of staying warmer than the surrounding environment. As was mentioned above, how intensively a colony heats its nest varies with whether or not the colony is rearing brood. If it is, then the central broodnest region is heated to 32–36°C. If the colony lacks brood, then the bees "turn down their thermostats," maintaining the core and mantle of the clustered bees only above about 18°C and 10°C, respectively. These two temperatures are critical minima: bees chilled below about 18°C cannot generate the action potentials needed to activate their flight muscles, their source of heat (Allen 1959a, Esch and Bastian 1968), and bees cooled below about 10°C become immobilized, entering a sort of chill coma (Free and Spencer-Booth 1960). Survival of such hypothermia depends on its duration; chilling to 10°C or colder kills most bees within 48 hours.

Colonies maintain a warm microclimate inside their nests through integrated controls on both heat loss and heat production. One key to understanding nest heating by honeybees is recognizing that a colony's emphasis on reducing heat loss or increasing heat production changes in a regular pattern as a function of ambient temperature, as will be discussed below.

Honeybees employ two general methods of improving heat retention in their nests. One, a long-term measure, involves selecting a protective nest site which will tightly enclose the colony's combs and bees (see Chapter 6). The size of the nest cavity's entrance opening is particularly important. Bees avoid cavities with apertures larger than about 60 cm^2, at least in part for reasons of energy conservation. Anderson (1948) observed that an uninhabited hive heated inside by a 15-W light bulb (comparable in heat production to a bee colony in mid-winter; see Chapter 4) was 3–8°C warmer on a winter afternoon when its entrance area was 9 cm^2 instead of 81 cm^2. The bees further minimize the draftiness of their homes by sealing unnecessary openings with sticky plant gums and resins.

The second general mechanism for reducing heat loss is the process of clustering, whereby the thousands of bees in a colony press tightly into a compact, roughly spherical, mass. Clustering begins when the temperature inside the nest dips below about 18°C. If the temperature outside the cluster falls further, the bees bunch closer together, thereby further shrinking the total cluster size, though at about − 10°C cluster contraction reaches its limit (Phillips and Demuth 1914, Wilson and Milum 1927, Free and Spencer-Booth 1958, Kronenberg and Heller 1982). Between 18 and − 10°C, the volume of a cluster shrinks roughly fivefold (Owens 1971). If a bee colony is killed in winter with hydrogen cyanide gas and the dead cluster is dissected, a two-part internal organization is revealed (Fig. 8.2). There is an outer zone consisting of several layers of densely packed bees sitting with their heads pointed inward. These workers form a mantle of insulation. They fill all the empty cells in combs and press together as closely as possible in the spaces between combs. Bees in the inner zone, in contrast, have room to crawl about, feed on the honey stores, fan with their wings, and tend brood (Gates 1914, Phillips

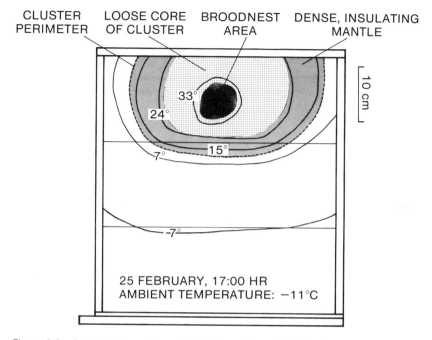

Figure 8.2   Anatomy of a winter cluster of honeybees. (Modified from Owens 1971.)

and Demuth 1914, Owens 1971). As discussed earlier, clustering reduces a colony's heat loss by shrinking its surface area for heat exchange. It also helps minimize heat loss by convection since by pressing together the bees reduce the porosity of their colony and therefore its internal convection currents. Perhaps most importantly, the dense outer layer of a cluster forms an effective blanket of insulation. Measurements of colony metabolic rate as a function of ambient temperature yield an estimate of heat conductance from winter clusters of just $0.10 \text{ W} \cdot \text{kg}^{-1} \cdot {}^\circ\text{C}^{-1}$ (Southwick and Mugaas 1971), a value which falls far below those calculated for reptiles, and rivals those of birds and mammals with their feathers and fur (Herreid and Kessel 1967).

These measures for heat retention are complemented in the honeybee's overall system of colonial thermoregulation by the capacity for heat production. A respectable amount of heat is generated by the resting metabolism of brood and adults—about 8 and 20 W/kg, respectively, at T_A of 35°C (Allen 1959b, Kosmin et al. 1932, Cahill and Lustick 1976, Kronenberg and Heller 1982). However, by revving up their flight muscles, adult bees can boost their metabolic rate and heat production to impressive heights, up to about 500 W/kg (Jongbloed and Wiersma 1934, Bastian and Esch 1970, Heinrich

1980). It is the adult bees' metabolic rate, therefore, which is adjusted to regulate a colony's heat production. Workers engaged in heat production for the colony are behaviorally indistinguishable from resting bees; both stand motionless on the combs. Recordings of thoracic temperature, measured with chronically implanted thermocouples, however, reveal striking differences between bees in the two roles. In "unemployed" bees, T_{Th} simply matches T_A. But in heat-producing bees, T_{Th} makes periodic, 1- to 5-min-long jumps, rising above T_A by up to 10° C, sometimes reaching 38°C before being allowed to decline as heat in the thorax radiates into the surroundings (Esch 1960) (Fig. 8.1). Measurements on single bees or small groups of bees confined in a controlled-temperature respirometer demonstrate clearly that worker bees will resist chilling by dramatically raising their metabolic rate (Free and Spencer-Booth 1958, Heussner and Roth 1963, Heussner and Stussi 1964, Stussi 1972). Thus whereas a group of 10 workers at T_A of 35°C showed little elevation in T_{Th} (36°C) and a low metabolic rate (29 W/kg), bees in a group held at 5°C boosted their metabolism to 300 W/kg and so maintained T_{Th} at 29°C, far above the ambient temperature (Cahill and Lustick 1976).

In nature, where each individual works together with thousands of colony members in resisting cold, this process of increasing heat production operates

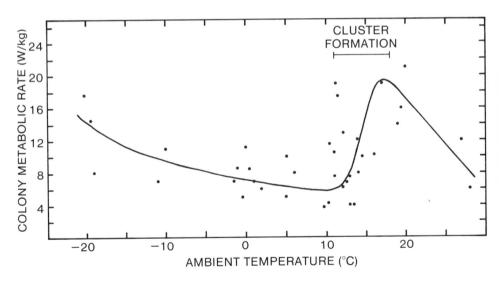

Figure 8.3 Colony metabolic rate as a function of ambient temperature. Each point represents the minimum metabolic rate recorded over a 24-hour period at one test temperature. (Modified from Southwick 1982.)

jointly with reducing heat loss by clustering. Figure 8.3 illustrates the broad pattern of coordinated interaction between these two thermoregulatory mechanisms. Heat production increases as T_A drops from 30°C to 18°C, and again when T_A falls below about 10°C, but decreases when T_A declines from about 18° to 10°C. Over this latter temperature range a colony coalesces into a well-insulated cluster and evidently can resist the deepening cold entirely by reducing heat loss. Because clusters continue to shrink down to about -10°C, improving heat retention evidently continues to play a role in colonial thermoregulation down to this temperature, though below 10°C it is joined by rising heat production (Free and Simpson 1963, Southwick 1982, Kronenberg and Heller 1982). Presumably the reason colonies do not initiate cluster formation at higher temperatures, and thereby reduce their nest-heating costs, is that coalescing into a tight cluster disrupts other colony operations, such as foraging, nest cleaning, and food storage.

The adaptive significance of nest heating—a means of speeding the development and raising the disease resistance of brood—is underscored by studies indicating that thermoregulation by bees depends not only on their own temperature, but also on that of the brood. When an aluminum plate beneath a layer of brood-filled cells was chilled from 30° to 10°C, while the air above the brood was maintained at 30°C, the bees tending the brood elevated their metabolic rate from 29 to 90 W/kg, thereby stabilizing the brood temperature at 30°C (Kronenberg and Heller 1982). Temperature receptors coating the five distal segments of workers' antennae (Lacher 1964) enable bees to monitor brood temperature closely, probably with a precision as fine as 0.25°C (Heran 1952). Worker bees evidently recognize cells containing brood by the associated odor of glyceryl–1,2–dioleate–3–palmitate (Koeniger and Veith 1983).

When colonies are without brood, and thus their warming activities serve entirely for the survival of the adult bees, it is less clear exactly what triggers thermoregulatory responses by individual bees. Presumably the bees in a cluster's mantle gauge just the temperature of their bodies or the surrounding air and either expand or contract this layer of insulation as needed. More problematic is whether the inner bees in a cluster, the ones warm enough to generate heat, adjust their heat production according to the needs of the outermost bees, which are frequently too cold to warm themselves, or whether the inner bees simply generate enough heat to keep themselves above about 18°C and this automatically provides sufficient warmth to keep the outermost bees always above about 10°C, the honeybee's minimum tolerable temperature. This matter will be taken up when examining temperature control in swarms, where the interactions between core and mantle bees have been analyzed in some detail.

Cooling the Nest

Although the rate of heat production by brood and nonincubating adults is low relative to that of an actively incubating worker bee, they do generate enough warmth so that at ambient temperatures above about 30°C, honeybee colonies face the problem of their broodnests overheating. As discussed above, the upper limit for broodnest temperature is about 36°C, with long-term temperature excesses of just 2–3°C severely disrupting brood undergoing metamorphosis. It is therefore not surprising that honeybees can cool their nests as impressively as they can heat them. When Lindauer (1954) placed a hive of bees in full sunlight on a lava field near Salerno, southern Italy, the colony's maximum internal temperature never exceeded 36°C, even though the outside temperature rose to 60°C and the interior temperature of a nearby, unoccupied hive climbed to 41°C. To counter overheating, colonies employ several mechanisms of nest cooling in a graded response, starting with simple spreading out of the adults in the nest and proceeding through fanning, water evaporation, and finally partial evacuation of the nest.

Adult bee dispersal within the nest at increasing T_A is probably simply an extension of the cluster expansion that starts when T_A rises above about $-10°C$. The precise environmental temperature at which this cluster expansion begins to be supplemented by fanning is probably quite variable and depends upon such factors as the nest's sun exposure, its insulation, and colony strength. Hess (1926) and Wohlgemuth (1957) report fanning when the nest temperature reached 36°C if not before. The fanning bees deploy themselves throughout the broodnest, aligning themselves in chains to drive air along existing (unidirectional) currents. Gangs of fanners face inward at the nest entrance and expel currents of warm air from the nest. Without entrance fanners there would probably be little exchange between the nest atmosphere and outside air through the small entrance opening.

High rates of airflow through the nest are created by these fanners. Hazelhoff (1954) constructed a hive with two openings, one at the top connected to an anemometer, and one at the bottom for the hive's entrance. Using this hive he estimated the flow rates produced by fanning bees. Once, when there were 12 strongly fanning bees spaced evenly across the 25-cm-wide entrance, the rate of air flow inward through the top entrance was 0.8–1.0 L/sec. Fanning bees use a markedly different pattern of wing movements than do flying bees, one which is designed to move air horizontally rather than to generate lift. The plane of wing movements, wingbeat frequency, and angle of attack all differ in the two behaviors (Neuhaus and Wohlgemuth 1960).

When cluster expansion and ventilation cannot cool the nest sufficiently, the powerful cooling technique of water evaporation is also brought into play (Fig. 8.4). The potential for nest cooling afforded by water evaporation is

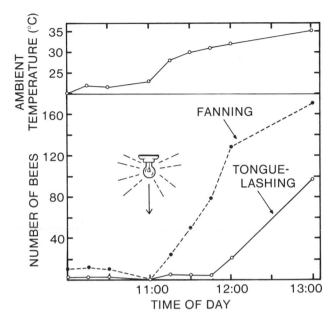

Figure 8.4   The number of honeybees fanning and tongue-lashing (drawing a fluid droplet out into a film with their tongue for rapid evaporation) as a function of ambient temperature, and after the colony was artificially heated. (From Lindauer 1954.)

demonstrated by the observations of Chadwick (1931) in California. One day in June when the midday air temperature rose to 48°C, the bees brought large amounts of water into their nests and there was little melting of combs. By 21:00 hours the temperature had dropped to 29.5°C, but at midnight a hot breeze from the desert raised the air temperature to 38°C. The supplies of water in colonies became exhausted, no more water could be collected until daylight, and some of the wax combs softened and collapsed.

The precise handling of water and the regulation of its collection formed the subject of intensive studies by Lindauer (1954) and Kiechle (1961). Water for nest cooling is distributed about the nest as small puddles in depressions on capped cells, smeared as a thin layer over the roofs of open cells, or placed as hanging droplets in these cells. It may also be rapidly evaporated via "tongue-lashing" whereby bees hang over the brood cells and steadily extend their proboscises back and forth. Each time a bee does this it presses a drop of water from its mouth and spreads the droplet with the proboscis into a film which has a large surface for evaporation.

A remarkable communication system is employed to regulate a colony's water collection. The fundamental complication in this operation is that the bees that sense the overheating and become "water sprinklers" are the young nurse bees in the central broodnest and not the older forager bees which can gather water. This division of labor was demonstrated by heating a small, central area of comb in an observation hive with a narrow beam of intense light and observing that this stimulated water collection by the colony's foragers without their contacting this region of the nest (Lindauer 1954).

How are the foragers informed of the need for cooling water? First, Lindauer (1954) found that colonies collect some water each day from spring to autumn for use in diluting honey to make brood food, if not for cooling of the nest. Thus there are almost always a few individuals foraging for water. To transmit the information whether or not the colony needs more water, the hive bees make use of the moment when they contact the water collectors during water delivery near the entrance hole. When a water shortage exists, a homecoming water forager is met at the entrance by hive bees which rush up to her and quickly suck up her extruded water droplet. Such a stormy reception informs the water forager that there is a pressing need for more water. On the other hand, when the overheating begins to subside, the hive bees show less interest in the water foragers. The latter now have to run around the hive themselves, trying to find a hive bee that will accept their load of water, and the delivery time increases. The delivery time appears to be an accurate gauge of a colony's water needs, though other actions of the hive bees toward foragers could also transmit this information. With delivery times of up to 60 sec, water collection continues, but if foragers cannot unload their water in 60 sec, their eagerness for collecting decreases rapidly. Water collecting almost disappears when delivery takes longer than 180 sec. A second point concerning recruitment is that when delivery times are very short (up to about 40 sec), the water collectors even perform recruitment dances after each collecting flight, and this stimulates other foragers to collect from the water source.

Finally, under conditions of extremely high temperature and high humidity, the bees may partially evacuate their nest and form a mass of hanging, clustered bees just outside the nest entrance. This reduces the heat production inside the nest and facilitates nest ventilation. Dunham (1931) observed the start of this when the coolest region of a colony's broodnest reached more than 34°C, and presumably fanning and evaporative cooling were already being applied in full force.

Thermoregulation in Swarms

The total resources possessed by a honeybee swarm as it sets out to establish a new colony are the bees themselves and the load of energy-rich honey carried inside each bee. Both of these assets are carefully conserved throughout the process of finding a new home. One expression of this resource conser-

vation is the remarkable docility of bees when in a swarm. So gentle are they
that a beekeeper can perform the seemingly amazing stunt of having a swarm
cluster on his chin, thus creating a "bee beard," with little fear of being
stung. By refraining from stinging unless severely disturbed, swarm bees help
preserve their fledgling colony's labor force. A second major mechanism for
preserving the swarm's capital is through energy-efficient thermoregulation,
recently analyzed in detail by Heinrich (1981a, 1981c). As is shown in Figure
8.5, the outermost bees in a swarm cluster are never allowed to cool below
about 15°C, and thus are protected from death due to chilling. Simultaneously,
the bees in a swarm's core maintain a 30–40°C microclimate deep inside the

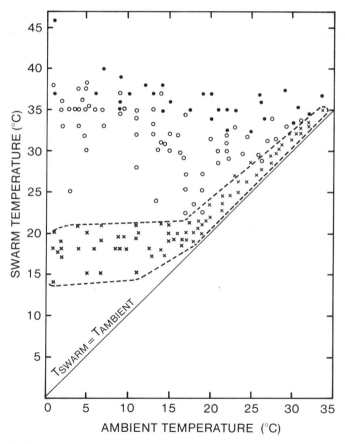

Figure 8.5   Core (circles) and mantle (crosses) temperatures of honeybee swarms as
a function of ambient temperature. Open circles: swarms of 2000–10,000 bees. Filled
circles: swarms of 15,000–17,000 bees. (From Heinrich 1981a.)

swarm. This pocket of broodnest-level warmth may help keep the swarm's queen and nurse bees physiologically primed for the massive brood rearing ahead. It may also facilitate the scout bees' flights from the swarm and so speed the passage through this perilous phase of the colony life cycle. The cost of this temperature control is astonishingly low. Metabolic rate measurements on whole swarms held at ambient temperatures of 0–35°C yield values generally in the range of 2–12 W/kg, thus well below the 20 W/kg metabolic rate of workers when resting (flight muscles inactive) at 35°C.

How do swarms obtain these elevated temperatures so cheaply? They do so primarily by skillfully trapping inside the swarm cluster the metabolic heat generated by the thousands of resting, immobile bees. This may seem surprising since swarms simply hang from tree branches, without the protection of a nest cavity, and so appear especially prone to heat loss. However, because a swarm has no brood or combs needing cover, its size, form and internal organization are all adjustable, and can be fully adapted for temperature control.

So effective are swarms at retaining their own heat that thermoregulation by bees in the core consists almost exclusively of adjusting the expulsion rate of *excess* heat. This is graphically illustrated by some of Heinrich's quantitative findings. For example, he found that if a swarm's core temperature (T_C) is 35°C, and T_A is 5°C, then the passive cooling rate of bees in the heart of an average-size, 12,000-bee swarm is only about 0.06°C/min. With this rate of cooling, a core bee seeking to maintain her body temperature at 35°C needs to expend energy only at a rate of 3.2 W/kg, which is only about one-sixth the resting metabolic rate of a bee at 35°C. Core bees rid themselves of their excess heat by forming open ventilation channels through the swarm and fanning to force convectional cooling. Only in tiny swarms, those with fewer than about 1000 bees, living at near-freezing temperatures, do the central bees in a swarm need to raise their heat production above resting levels to maintain a 35°C microenvironment.

The bees forming the mantle of a swarm cluster experience far poorer insulation than do the bees buried deep inside, and therefore are not always oversupplied with heat. Thus, to give an extreme example, at T_A of 5°C, the cooling rate of a bee in a swarm's outermost layer is about 0.20°C/min, a rate which implies that maintaining a body temperature of 18°C requires a metabolic rate of roughly 11 W/kg. This is probably far above the resting metabolism of a bee cooled to 18°C. (Measurements of resting metabolic rate as a function of temperature are not available for honeybees. However, if the pattern for honeybees resembles that for bumblebees (Kammer and Heinrich 1974), then given a resting metabolic rate at 35°C of 20 W/kg, it appears that resting honeybees at 18°C are expending energy at a rate of only about 2 W/kg.) Although mantle bees must sometimes increase their heat production, before doing so they take several steps to minimize heat loss from the swarm and thereby postpone the point of activating their flight muscles. By pressing in toward the swarm's center they can contract the cluster's surface area, diminish the mantle's porosity, and fill the interior ventilation channels. Overall, in a transition of T_A from 30°C down to 5°C, the density of bees in a swarm cluster increases from about 0.13 to 0.50 bees/cm³. As a result, the mantle bees' own rate of heat loss, as well as that of the core bees, decreases and so the swarm's precious energy supply is conserved.

What is perhaps most surprising about temperature control in swarms, and perhaps also in broodless winter clusters, is that this fundamentally social process seems to operate without any communication between colony members. Each bee acts independently, responding to her own particular thermal environment. Thus bees in the core, which are well insulated against heat loss, are heated passively by their resting metabolism, while bees in the mantle, with less insulation, must shiver to stay warm, but do so only when cooled to about 15°C. Hence despite a lack of central coordination, clusters of bees achieve highly economical temperature control.

Thermoregulation during Foraging

The honeybee's ability to thrive in a vast range of environments, from temperate regions to humid tropical and hot desert habitats, reflects powerful temperature control not only by whole colonies inside their nests, but also by individual foragers out amidst the flowers. These bees must maintain their body temperature below the lethal upper limit of 45–50°C, and, while flying, must keep their thorax temperature above 27°C, the lower limit for steady flight (Heinrich 1979b). In ways which are just beginning to be understood, foraging bees manage to fulfill these requirements across the range of ambient temperatures from about 5° to over 45°C, thus achieving considerable freedom from temperature limitations when collecting resources for the colony.

It is now well established that the heat for elevated thoracic temperatures at low T_A comes from the bees' flight muscles and that bees do not regulate heat production while in flight (Esch 1960, Esch and Bastian 1968, Heinrich 1979b). Given the size of a worker bee, its flight speed, degree of insulation, and other factors affecting the passive cooling rate of a bee, the heat generated during flight is enough to maintain a bee's T_{Th} above T_A by about 15°C. Thus a bee should be able to keep her thorax hot enough for flight (above 27°C) at ambient temperatures down to about 12°C (Heinrich 1979b). In fact, when bees were forced to fly in a temperature-controlled room at a slightly lower temperature, 10°C, they could not maintain continuous flight, but instead had to alight with chilled flight muscles after 30–120 seconds. However, in nature, bees will fly out of their nests and forage at temperatures well below 12°C. Heinrich recorded foragers returning to their hive with an average T_{Th} of 30°C when T_A was just 7°C. Exactly how bees accomplish this remains unclear. Their solution may involve periodic landings for warm-up, with the bees taking advantage of lower cooling rates when still as compared with the higher rates found under the wind-chill influence of flight. Heinrich's bees which were forced to land by flight-muscle hypothermia did sometimes then raise T_{Th} up to 39°C and again launch into flight. Such stop-and-go foraging is expensive in time and energy spent on warm-up, but can still be profitable when working rich flowers, especially since once a bee reaches the flowers her behavior will likely consist of intermittent flight between flowers, thus automatically affording her opportunity for warm-up when perched on the flowers.

One element of honeybee anatomy which helps them maintain a high T_{Th} at low T_A is the highly convoluted design of the aorta as it passes through the petiole (anatomy described by Snodgrass 1956, Wille 1958). This prolongs the passage of cool blood flowing forward from the abdomen into the thorax, thus increasing its recovery of heat by counter-current exchange from warm blood passing rearward into the abdomen (Heinrich 1980). Although invaluable at low T_A, this structure prevents honeybees from using their abdomen as a heat radiator at high T_A, the main cooling mechanism for other endothermic insects such as bumblebees and dragonflies (reviewed by Kammer 1981). Nevertheless, honeybees do possess an effective cooling system for flight at high ambient temperatures. Thus whereas T_{Th} of a flying bee exceeds T_A by 15°C when T_A is low (15–25°C), when T_A is high (45°C), T_{Th} just matches T_A. The primary mechanism for cooling the thorax appears to be to keep the head cool by regurgitating fluid from the honeystomach and holding the extruded nectar droplet between the folded tongue and mandibles (Heinrich 1980). Water evaporation from this droplet keeps the head temperature (T_H) below T_{Th} by 2–3°C on average. The head therefore functions as a heat sink for excess thoracic heat, the overall result being that holding $T_H < T_{Th}$ maintains $T_{Th} = T_A$. Thus honeybees, though they produce prodigious quantities of heat while flying, can fly and forage at ambient temperatures as high as the lethal upper limit for resting bees.

9 Colony Defense

Evolutionary Perspectives

The honeybee's defenses against predators and parasites are awesomely diverse. There exist the expected mechanisms of defense: protective nest sites, guard bees at the nest entrance, venomous stings, disposal of diseased brood, and massive counterattacks synchronized by alarm pheromones. But to this list we must add other, often subtle, and sometimes even bizarre techniques: specialist "undertaker" bees, enzymatic generation of hydrogen peroxide in ripening honey, varnishing of the nest interior with fungicidal and bactericidal plant resins, filtration of disease spores from food, genetic systems for discriminating nestmates from outsiders, and still others.

One approach to gaining perspective on this intricate subject is to examine its evolutionary background. The honeybee's multitude of defense traits reflects in part the ecological reality that all organisms face a legion of predators and parasites, each equipped with a unique array of tools for disarming its prey or host (Janzen 1981). However, the wealth of honeybee defense adaptations also reflects certain distinctive properties of this insect's biology. One is the sessile nature of honeybee colonies. Because their nests are costly to build, and are often filled with expensive brood and essential stored food, colonies usually cannot afford to evade an attacker by simply running away. Instead, they must try to repel invaders in standing battles, employing a wide array of biochemical, morphological, and behavioral weapons.

A second property of honeybee colonies which makes them an especially attractive target for predation and parasitism is simply the wealth of resources stockpiled inside their nests. Usually a nest contains several kilograms of nutritious brood and honey, powerful attractions for a multitude of animals. Moreover, the warmth and high humidity inside honeybee nests result in an almost perfect environment for the incubation of fungi and bacteria, at least on the nest's margins, away from the very high broodnest temperatures. The traits which make honeybee nests bonanzas for predators hence stem primarily from the honeybee's sociality. Thus here we have the paradoxical situation that although improved nest defense was probably the primary ecological factor favoring the evolutionary origins of bee societies (Michener 1974, Evans 1977, see Chapter 3), later advances in bee sociality have actually intensified the problem of nest defense. In all, there are at least several hundred organisms, ranging from viruses to vertebrates (reviewed by Morse 1978), which would consume some part or all of a honeybee colony if they could penetrate the bees' defenses.

Colony defense is a rich topic of study, but one with special difficulties. These derive primarily from the fact that defense effects are not intrinsic to a defender's biology, but only exist in the context of the invader's means of assault. For example, the honeybee's sting provides potent defense against birds and mammals, but is powerless against mites or bacteria. Likewise, the high sugar concentration of honey provides thorough protection against consumption by yeasts (the high osmotic pressure kills yeast spores), but only heightens its attractiveness to sweet-toothed vertebrates. Therefore, understanding colony defense involves analyses of predator-prey or parasite-host interactions; one needs the presence of both attacker and defender to understand fully the honeybee's mechanisms of self-preservation. This may prove impossible for certain protective measures, either because the predator which shaped the defense mechanism has gone extinct sometime during the honeybee's 35-million-year history, or because improvements in the bee's defenses have caused the predator to cease preying on honeybees. Both processes produce defense mechanisms which are anachronisms and thus possess uninterpretable natural histories. Even if the predator still exists and continues to prey on honeybees, it may exist at historically low densities, owing to human disturbance of habitats. The clear message here is that studies of

Figure 9.1   A guard bee on alert at her nest's entrance.

honeybee defense biology, especially experimental analyses of the importance of particular defense traits, must be conducted in habitats that have suffered minimal destruction.

A second complication to the analysis of defense traits derives from their diversity and integration with other aspects of honeybee biology. Many traits which evidently contribute to colony defense actually confer multiple benefits to colonies. A case in point is the small entrance opening of honeybee nests, a product of the bee's behavior during nest-site selection (see Chapter 6). A small nest aperture certainly helps bees keep out predators, but it also helps them control the nest microclimate. Such situations complicate the puzzle of the importance of predation in shaping a trait. Where predation is but one of several selective forces, the trait is likely to reflect a compromise among competing design factors, and so may possess shortcomings when viewed purely as a protective device. Thus, as will be discussed later in the chapter, multiple functions of traits involved in colony defense greatly complicate the cost-benefit analyses of an organism's defenses.

Catalogue of Protective Mechanisms

The various physiological, morphological, and behavioral devices through which workers shield their colony from destruction involve nearly all aspects of honeybee biology. To examine this generalization properly, let us now review those traits that can be identified as the principal lines of colony defense.

Protective nest site. A colony's scout bees make a major first step toward colony safety in selecting a nest cavity with a small (less than 60 cm^2), high (more than 3 meters above ground) entrance opening (Seeley and Morse 1976, 1978). Such nest sites are relatively difficult for large, visually orienting predators to discover, approach, and penetrate. Although the matter has never been investigated, it seems likely that scout bees also favor nest sites with sturdy cavity walls and possibly even inconspicuous entrances.

Guard bees. When workers make the transition from hive bee to forager (age caste III to age caste IV, see Chapter 3), approximately 20 percent of this population will spend a day or two serving as guards at the nest entrance (Lindauer 1952, Sekiguchi and Sakagami 1966). Here they adopt a charac-teristic stance—forelegs off the substrate, antennae projected forward, and sometimes wings and mandibles spread—ready to rush toward any intruder (Fig. 9.1). Guards intercept and examine with their antennae bees entering the nest, such examinations generally lasting just 1 to 3 seconds. This evidently provides sufficient time to distinguish nestmates from robbers coming from other colonies (Butler and Free 1952).

Guards thwart robber bees by clamping onto a leg or wing with their mandibles, and then, once a firm hold is planted, curling the abdomen under to bury a sting in the thief. Intruding wasps and other flying insects are likewise blocked from entering the nest. To repel an intruding ant, a guard will run toward it, then pivot 180° when directly in front of the ant, simultaneously fanning her wings and often kicking her hindlegs rearward to strike the ant. Together these behaviors dislodge the ant from around the nest entrance (Spangler and Taber 1970). Vertebrate invaders are perceived by guards as dark, rough-textured, objects which move and carry an animal scent (Free 1961). Guards respond to such a stimulus by launching out at the intruder, burrowing deep into its fur while producing a harsh buzzing sound with their wings, and implanting their stings.

Nestmate recognition. During a dearth in forage, a colony's foragers frequently attempt to rob neighboring colonies of their honey stores. Thus one of the more common, and probably the most difficult, type of intruder detection performed by guard bees involves discriminating nestmate and non-nestmate workers at the nest entrance. Robbers frequently adopt a tell-tale, jerky flight pattern as they try to slip past the guards, and guards evidently use this as a visual sign of a robber bee (Butler and Free 1952). However, such hesitating flight seems to follow having been caught by a guard, so the ultimate basis of robber identification probably involves odors (Ribbands 1954). The behavior of guards when examining suspect bees—they closely approach and touch the bees with their antennae—itself suggests that workers carry colony-specific odors. The first experimental proof of colony odors came from studies in which bees from two colonies were trained to forage from two distinct, but closely spaced, feeders offering identical food, and recruits appeared predominantly at the feeder worked by their nestmates (von Frisch and Rösch 1926, Kalmus and Ribbands 1952). Although undoubtedly resulting from a complex blend of scents, each forager's colony odor is not masked or modified by the additional scents she acquires while foraging. When Butler and Free (1952) captured pollen-bearing bees returning to their hives, and placed them at the entrance of their own or a foreign hive, they found marked differences in the probability of their being accepted: 0.98 and 0.37, respectively.

These colony-specific odors may have both environmental and genetic origins. The ability of a colony's guards to recognize robbers can be experimentally increased by feeding colonies distinctly scented foods, such as black molasses or heather honey (Kalmus and Ribbands 1952, Köhler 1955). Thus it appears that in nature, colonies acquire distinctive odors by collecting different blends of nectar and pollen from the array of available flowers. These floral odors are then presumably adsorbed onto the waxy epicuticle of each bee in the colony. The possibility of genetic influence on colony odor derives support from studies by Breed (1983) which showed that when worker pupae were removed from their nest and reared in an incubator (thus minimizing their exposure to odors in their colony's nest), and then placed at the entrance of a hive shortly after emergence, the probability of these workers being attacked by guards was significantly lower when placed at the entrance of their maternal colony than at that of an unrelated colony (probabilities of 0.28 and 0.73, respectively). One conclusion that could be drawn from this experiment is that young bees secrete genetically controlled odors which guard bees can use to recognize colony members. How long such endogenous odors remain important in the life of a worker bee, their glandular origins, and their role relative to environmental odors in forming a colony's complete odor, are all mysteries. Likewise, the chemical nature of colony odor and the details of when and how individuals learn their colony's odor stand as important puzzles for future research.

Alarm-recruitment pheromones. When a bee guarding her nest's entrance is struck by an intruder, she raises her abdomen and protrudes her sting, whereupon the surrounding bees instantly go on alert, either standing ready for attack with wings and jaws spread wide, or launching into flight in search of the foe. If severely disturbed, the guard will retreat inside the nest, there exciting additional guards to join in the fight. For example, when Maschwitz (1964) monitored a colony one evening for 24 minutes without creating any disturbance, he observed just one bee patrolling the entrance opening, but when he pinched 8 bees in the hive's entrance, 140 guards boiled out in an aggressive frenzy.

The alarm-recruitment signal inciting these mass defensive attacks is a pungent odor produced primarily by cells lining the sting chamber. Chemical analysis of the volatile compounds extracted from the sting apparatus of guard-age workers reveals about 70 percent isopentyl acetate (also found in banana oil) with the remainder a complex blend of mostly C_4 to C_9 alcohols and their acetates (Boch et al. 1962, Boch and Shearer 1966, Blum et al. 1978, Collins and Blum 1982, 1983). A second source of alarm pheromone, the mandibular glands in the head, produces 2-heptanone, though this compound appears to be at least 20 times less potent than isopentyl acetate in signaling alarm (Shearer and Boch 1965, Boch et al. 1970). Guards often grip with their mandibles objects they are stinging, so 2-heptanone is probably daubed on enemies during attack. The sting pheromone system operates even more effectively as a recruitment signal to an attack site. Because honeybees' stings are barbed, their stings and adjacent tissues remain anchored in the skin of intruders, thereby creating a chemical beacon which guides other guards to the enemy. Beekeepers have long recognized that once one bee has sunk her sting in exposed skin, her nestmates will quickly join the attack, often implanting their stings within a few centimeters of the first.

Venomous sting. The honeybee's poison-bearing sting is exquisitely designed for inflicting pain or death on predators. To understand the effectiveness of this means of defense, we must consider the details of its morphology and biochemistry. The entire sting apparatus (Fig. 9.2) consists of two sets of functionally distinct parts: the motor apparatus, an assembly of plates and muscles which powers the sting, and the long tapering shaft, which pierces the foe's skin and conducts venom to the wound (Snodgrass 1956, Maschwitz and Kloft 1971). When a bee stings, the initial insertion of the sting results from a quick jab of the abdomen, but the subsequent deeper penetration results

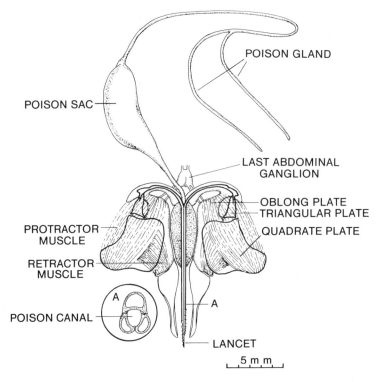

Figure 9.2 The sting mechanism of worker honeybees, showing the moving parts (lancets and articulated plates) in thick lines. The small drawing on the left shows a cross-section through the sting's shaft. (Modified from Dade 1977.)

from successive alternating movements of the two lancets forming the sting's sharp shaft. After each thrust, each lancet holds the sting in place with the recurved lateral barbs on its distal end, while the other lancet overreaches the first and pulls the sting into a deeper position. Each lancet is flexible and runs along a semicircular track at the proximal end, and is thrust in and out by alternating contractions of the retractor and protractor muscles attached to its lever-like oblong, quadrate, and triangular plates, as is shown in Figure 9.2. An oscillator circuit in the last abdominal ganglion times the alternating muscle contractions. As the sting digs itself deeper and deeper into the predator's skin, venom from the poison sac is simultaneously driven through the poison canal and out the tip of the sting. The entire sting apparatus has only a delicate membranous connection with the surrounding walls of the sting chamber, thus a light pull on the sting tears it from the living bee. The complete apparatus—venom sac, motor apparatus, sting shaft, and ganglion—is extracted and continues pumping poison automatically, thus delivering a far larger dose of venom than the bee could otherwise deliver before the predator brushes her off.

The intricate anatomy and efficiency of the sting apparatus is complemented by the complex biochemistry of bee venom (reviewed by Habermann 1971, 1972, O'Connor and Peck 1978, Riches 1982). Its constituents can be categorized as biogenic amines, toxic polypeptides, and enzymes. The low molecular weight amines—such as histamine, dopamine, and noradrenaline—represent less than 2 to 3 percent of venom and are overshadowed by the compounds in the two groups of high molecular weight agents. The toxic polypeptides include melittin, apamin, and a mast cell degranulating peptide. The last two substances cause the destruction of neurons and massive release of histamine, and therefore are potent toxins, but together produce far less destruction than melittin, which constitutes 50 percent of dry venom. Because this polypeptide bears a string of hydrophilic and hydrophobic side chains at opposite ends, it functions as a detergent, a "structural poison," disrupting membranes and other lipid-water interfaces. Melittin has proven destructive in every pharmacological setting studied so far, including the lysing of erythrocytes, mast cells, and leukocytes and their lysosomes, and dilating blood vessels. It also destroys enzyme systems bound to membranes, thereby diminishing electron transport in mitochondria and uncoupling oxidation from phosphorylation. In contrast to melittin, which alters tissue structure by changing its physicochemical properties, the enzymes in bee venom disrupt cells by enzymatic hydrolysis. Hylauronidase seems to serve as a spreading agent, opening the way for other venom components by digesting connective tissue. Phospholipase A attacks structural phospholipids. Since these are integral parts of biological membranes, mitochondria, and other components of cells, their loss triggers cell failure.

Large colony size. With populations typically in the range of 10,000 to 40,000 bees, honeybee colonies are among the largest known for social bees (Michener 1974). This large colony size undoubtedly reflects several factors, including the energetics of temperature control and foraging, but certainly also colony defense. Because of their large size, honeybee colonies can escalate their defense responses to extremely high levels, with up to a thousand or more bees simultaneously burrowing into a predator's feathers or fur, penetrating its nostrils and ears, and implanting fiery stings.

Reproduction by swarming. As was discussed in Chapter 4, the honeybee's habit of colony reproduction by swarming probably traces ultimately to selection for superior defense of queens. Swarming achieves this by maintaining a contingent of guard workers around each queen throughout her life.

Synchronous orientation flights. When a colony's young bees perform their initial trips outside the nest, they present a remarkable sight. On warm, sunny afternoons, for periods lasting generally just 20 to 60 minutes, hundreds of workers pour from the nest and form a cloud of bees hovering about the nest's entrance (Sakagami 1953, Vollbehr 1975). Once outside, each bee lingers in the cloud only about 10 seconds before heading off on a short flight across the surrounding countryside, but upon her return she hovers facing the nest for a few minutes. In the course of these actions each bee memorizes the appearance of her nest's entrance opening and learns how to orient to and from her home over distances of several kilometers (von Frisch 1967). These orientation flights are performed by bees ranging in age from 4 to about 18 days. Synchrony is evidently achieved by the older bees, which work near the nest entrance and so can monitor the weather outside, signaling suitable times for orientation flights by scrambling through the nest and jostling their sisters. It seems likely that this synchronization has the beneficial effect of saturating predators at the nest entrance, such as lizards or toads, and so increasing the survival rate of workers venturing outside the nest.

Removal of dead bees. Honeybee colonies possess several adaptations for the rapid disposal of dead bees, which, if they accumulated in nests, would foster the spread of diseases and parasites within colonies. One such trait is the small group of workers, numbering 1–2 percent of a colony's population, specializing on the task of removing dead adults (Visscher 1983). These ''undertaker'' bees patrol the floor of their nest for corpses. Because dead bees rapidly undergo a change in odor, either gaining or losing some olfactory cue within 15 minutes of death, they are readily recognized and speedily removed. Thus, whereas loose bits of debris or resting, motionless live bees are ignored by the undertakers, dead bees are usually hauled out within an hour of death. Consequently, although the daily mortality rate inside a colony's nest is approximately 100 bees (Gary 1960), nests rarely contain more than one or two dead adults.

Colonies also quickly discard dead brood. This is important in resisting several brood diseases, including sacbrood and European foulbrood, but it has been studied most intensively with larvae killed by the highly contagious disease American foulbrood, caused by the bacterium *Bacillus larvae*. This bacillus usually infects bee larvae less than two days old, but cannot produce more infective spores until the larvae reach the age of 11 to 12 days, at which time some 2500 million spores per larva are formed (Bailey 1981). Resistance to the disease therefore requires early detection and removal of infected larvae. Colonies can achieve this, removing all diseased brood by day 12 of larval life (Woodrow and Holst 1942, Rothenbuhler 1964), thus just before they begin to fill with spores (see Fig. 9.4). After extracting the larvae from their cells, the workers either eat them (adults do not contract the disease) or fly out and drop them 10 to 100 meters from the nest.

Collection of plant resins (propolis). Propolis is the general name for the resinous material gathered by honeybees from various plants. The word derives from the Greek *pro-* (for or in defense), and *polis* (the city), thus meaning in defense of the city or colony. In north temperate regions, the principal sources of propolis are various species of poplar, elm, birch, alder, beech, horse chestnut, pine, and spruce trees. A few foragers from a colony visit the buds or wounds of these trees, where they tear off bits of the sticky resins and pack them onto their corbiculae (pollen baskets) for the flight home (Meyer 1956b). Bees use the propolis to plug cracks and holes in their nest's walls, reinforce their wax combs, reduce their nest's entrance opening to render it more weathertight and easier to defend, embalm the carcasses of invaders (such as mice or wax moths) which they have killed but cannot carry outside, and build a smooth, clean coating over the nest cavity's walls. When fresh, the tree resins are so sticky that propolis foragers require unloading by other bees, but as time passes they dry and turn brittle.

Besides filling the purely mechanical roles of a glue or space-filling cement, propolis also serves in colonial defense against fungi, bacteria, and viruses. This is hardly surprising, given that the plants' reason for synthesizing the resins collected by bees is to defend against predators, including microorganisms (Swain 1977, Harborne 1982). In essence, the bees are pirating chemical weapons from the trees. When Lindenfelser (1967) tested the antimicrobial activity of propolis he found strong inhibitory activity *in vitro* in 25 out of 39 bacterial species, most notably *Bacillus larvae*, and growth inhibition in 20 out of 39 fungal species. The compounds responsible for these antibiotic effects are just beginning to be identified chemically (reviewed by Ghisalberti 1979). The gross composition of propolis is roughly 70 percent resin (natural polymers), 25 percent beeswax, and 5 percent volatile oils. Over 30 compounds, mainly flavonoids, have been isolated from the volatile oil fraction, many of which exhibit antiseptic properties. For example, pterostilbene, which is a potent inhibitor of various fungal species (Lyr 1961), is a common constituent. So are ferulic and caffeic acids, which both possess antibacterial activity against *Staphylococcus aureus, Proteus vulgaris, Candida diphteriae*, and various other gram-positive and gram-negative bacteria, as well as antifungal activity against *Helminthosporium carbonum* (Čižmárik and Matel 1970, 1973). Other identified components with antimicrobial activity include the flavones galangin, chrysin, isalpinin, and tectochrysin, and the flavanones sakuranetin, pinobanksin, and pinocembrin (Villaneuva et al. 1964, 1970, Schneidweind et al. 1975). No doubt many more biologically active compounds will be identified from propolis. To place these materials in a meaningful perspective, future studies must determine which substances both inhibit microbes damaging to honeybees, and occur at sufficient levels to provide a potent defense.

Gut anatomy and defecation. The design of the honeybee's alimentary canal and defecation behavior further strengthen a colony's defenses by providing a mechanism for the collection and removal of spores of disease-causing microorganisms. The most intriguing component of this micro-cleaning system is the proventriculus organ, a combination valve and filter located between a bee's crop (food storage chamber) and ventriculus (digestive chamber). This organ efficiently scavenges particles suspended in the crop through the action of its hair-lined lips, which project into the crop and perform gulping movements, opening and closing so that particles are caught by the hairs while fluids return to the crop (Bailey 1952). The primary role of the proventriculus is extraction of pollen grains (typically 5–100 μm in diameter) from the crop, so that they can be concentrated in the ventriculus for digestion by proteolytic enzymes. But it is also effective in trapping smaller particles, such as the spores of disease agents like *Bacillus larvae* and *Nosema apis* (spore diameters of about 1 and 5 μm, respectively) (Bailey 1952, Sturtevant and Revell 1953, Thompson and Rothenbuhler 1957). For example, the number of *Nosema* spores in the crops of workers fed a concentrated solution of these particles declines logarithmically, halving roughly every five minutes (Fig. 9.3). *Bacillus larvae* spores, being somewhat smaller, are collected more slowly though still quite rapidly; within 40 minutes of ingestion, over 75 percent of the spores imbibed are sequestered behind the proventriculus (Wilson 1971). Thus when workers suck up fluids around larvae dead from American foulbrood, collect spore-laden honey from colonies killed by this disease, or chew up feces contaminated with *Nosema* spores, the disease spores that they ingest are quickly concentrated in their hindguts.

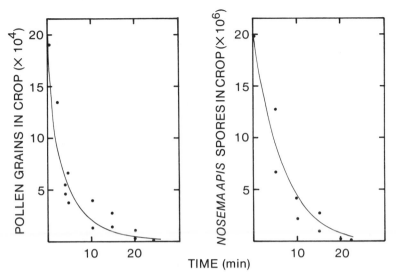

Figure 9.3 Pollen grains (left) and disease spores (right) are rapidly filtered out of a worker bee's crop through the scavenging action of the proventriculus organ. (From Bailey 1952.)

In defecating, workers complete the cleaning process. Almost without exception the bees eliminate outside the nest. Even in late winter and early spring, following several months of confinement and when flight opportunities are rare, the vast majority even fly out 10 to 100 meters from the nest before dropping their feces (Riemann 1958). The honeybee's highly expandable rectum, which in late winter nearly fills the abdomen and whose contents can constitute up to 40 percent of an individual's weight, plays a central role in giving bees precise control over the time and place of defecation (Lotmar 1951, Nitschmann 1957). The crucial importance of voiding outside the nest is illustrated by the pattern of outbreaks of *Nosema* disease. Caused by a protozoan which develops exclusively in the midguts of adult bees, this disease spreads from one generation to the next within a colony only when young bees find combs to clean which are contaminated with *Nosema*-infected feces. Thus about the only time the disease occurs is in late winter/early spring, following instances of defecation inside the nest by overwintering bees (Bailey 1981).

High broodnest temperatures. As discussed in Chapter 8, a colony's maintenance of its broodnest at 32–35°C evidently increases the resistance of its members to certain viral, fungal, and bacterial diseases.

Antibiotics. Perhaps the most poorly understood weapons in a honeybee colony's defense arsenal are the antibiotics which their workers synthesize. The sterility of guts of healthy larvae (Gilliam 1971), the inability of *Bacillus larvae* rods to multiply or sporulate in adult workers (Wilson 1971), the long-term storage of honey without spoilage (White 1975), and the many other striking features of the honeybee's microbiology (reviewed by Lavie 1968), indicate powerful defenses against microorganisms, but to date only two antibiotic systems have been identified biochemically.

As early as 1937, it was recognized that honey possesses antibiotic properties (Dold et al. 1937, Prica 1938). A major factor in this resistance to microorganisms is the high osmotic pressure in honey. When fully ripened by the bees, honey contains only 14–20 percent water and thus is an intensely hygroscopic environment in which bacteria and yeasts are rapidly dehydrated (White 1975). Freshly collected nectar, on the other hand, frequently contains 60 to 80 percent water (Southwick et al. 1981) and therefore constitutes a potentially ideal growth medium for microorganisms. In actuality, however, dilute honey solutions are even more bactericidal than is fully concentrated honey. The principal source of the non-osmotic antibiotic effect of honey is the hydrogen peroxide it contains, generated by the enzyme-catalyzed reaction of glucose with oxygen and water to form gluconic acid and hydrogen peroxide (White et al. 1963). The enzyme—glucose oxidase—is synthesized in the hypopharyngeal glands of workers (Gauhe 1940) and is probably mixed with nectar even as it is collected from flowers (D. M. Burgett, personal communication).

The second known antibiotic produced by honeybees is 10–hydroxy–2–decenoic acid, a fatty acid produced by the mandibular glands of nurse bees and added to the food mixture given to larvae (Butenandt and Rembold 1957, Callow et al. 1959). Here it acts as a general antibiotic against bacteria and fungi (McCleskey and Melampy 1934, Blum et al. 1959). One of its major effects is probably to raise the bees' resistance to *Bacillus larvae*. Brood food has been shown to delay the germination of *B. larvae* spores (Rose and Briggs 1969), and because the susceptibility of larvae to infection by this bacterium falls off dramatically between 5 and 35 hours after hatching (Bamrick and Rothenbuhler 1961, Bamrick 1967), even small delays in germination probably help inhibit the pathogenesis of the bacterium.

Cost-Benefit Analyses of Defense

Prior research on colony defense by honeybees consists primarily of detailed studies of how bees protect themselves from attackers, an approach which will remain important long into the future. Many basic questions about how predators and parasites are thwarted remain unanswered, including the full mechanism of nestmate recognition, the functional chemistry of propolis and venom, and the role of broodnest temperature in combating brood diseases. Moreover, because there is no single, global principle of design underlying the honeybee's defenses, deep understanding of this insect's protection system will come only from a detailed knowledge of its broad suite of defenses.

It is also clear, however, that future research must place greater emphasis on a second approach to colony defense, one which addresses the costs and benefits of particular defense traits. Here the goal is not greater understanding of how each weapon works, but why bees possess particular defense traits and why they make adjustments in the use of certain defenses. The questions at issue here resemble familiar military problems. For example, invasion is damaging, but defense can only be undertaken at cost. Which defenses are worthwhile and how much should be invested in each of the worthy ones? If we denote by $\bar{B}_i$ the average gain in fitness per generation resulting from possessing a particular defense trait i, and by $\bar{C}_i$ the average cost of having this trait measured as the decline in fitness resulting from diverting resources to this defense trait instead of to colony growth, food storage, reproduction, or other lines of defense, then defense by technique i should only occur if $\bar{B}_i > \bar{C}_i$. It is important to recognize that B_i and C_i are not constants, but change with time, co-varying with such things as colony size, food stores in the nest, the stage in the colony cycle, and density of predators. For example, the benefit of having 50 suicidal guard bees stationed at the nest entrance probably rises as the number of lizards or skunks living nearby increases, while the cost of this guard detachment falls as colony size increases. As a first approximation, we may expect that colonies try to maximize the quantity $(B_i - C_i)$ for each trait i at all times.

Doing so poses especially interesting problems for those defense traits in which bees can adjust their investment, and so regulate the benefits and costs of employing these lines of defense. Janzen (1979) draws the analogy here to a homeowner buying different amounts of fire insurance depending on his current neighborhood, the construction of his house, and the value of belongings in the home. Some of the honeybee's defenses are essentially fixed—such as its nest site and sting apparatus—and cannot be tuned to changes in circumstances, but most are adjustable. For example, the intensity of entrance guarding changes markedly across a summer. During times of rich forage, experimentally introduced intruders easily slip inside foreign nests, but at times of dearth these same bees are consistently rebuffed when they try to enter another colony's nest (Butler and Free 1952). Likewise, it is well known by beekeepers that the chance of getting stung when opening a beehive rises dramatically across a summer as the colony grows in population. The principal question is how finely do bees adjust their colony's defenses in relation to their circumstances? Finding the answers involves determining the cost and benefit patterns across time for particular defense traits. To do this, it must be understood that the relevant currency of benefit and cost is fitness—the number of reproductives reared to maturity—not honey production, pollen intake, comb construction, or brood rearing, unless these can be accurately translated into fitness effects. Such translation will prove tricky. For example, the benefit to a colony in late winter of having stored an additional kilogram of honey the previous summer will differ radically depending on whether the colony's honey stores are plentiful or running low.

One approach to conducting a cost-benefit analysis of a defense trait is to compare normal and genetically mutant colonies, thus ones with and without the defense. If the genetic difference is clean, involving only the trait under investigation, then differences between the colonies should reveal the economics of the trait. A preliminary example of this approach can be given for the defense behavior of removing larvae infected with *Bacillus larvae* (American foulbrood). The detailed studies by Rothenbuhler and his colleageus (reviewed by Rothenbuhler 1964) have demonstrated that colonies exist with and without this behavior, and that the difference is evidently traceable to two genes, one influencing the behavior of uncapping cells, the other the act of extracting the larvae from their cells. Curiously, this hygienic behavior is rarely expressed by bees, even though it greatly heightens a colony's resistance to American foulbrood. Each of the two control genes has a dominant allele which blocks the associated behavior, and, more importantly, the recessive alleles which allow the behavior to unfold evidently occur at low frequencies in natural populations.

This all suggests that natural selection has favored the spread of the dominant, behavior-blocking alleles, presumably because the costs of this defense trait exceed its benefits. Comparisons between colonies with and without the behavior (Fig. 9.4) suggest that this might be so and why. Colonies exhibiting the behavior probably suffer a major cost because their workers frequently uncap and remove even healthy larvae, and this is a cost they bear all the time, whether or not the colony is infected by *Bacillus larvae*. The countervailing benefit of the behavior, high resistance to American foulbrood, is only enjoyed when colonies are exposed to this disease. Thus it seems likely that unless the incidence of American foulbrood is high, the cost of hygienic behavior greatly exceeds its benefit.

A second method for analyzing the economics of a defense trait involves experimentally disrupting the defense and measuring the fitness effects of the ensuing damage. Because one is not changing the colony's properties, only breaking down one of its defenses, one will generally not detect the costs associated with the trait, only the benefits. An example of this approach comes from a study with *Apis florea* (discussed in Chapter 10), which positions its open-air nests amidst dense foliage for concealment from predators (Seeley

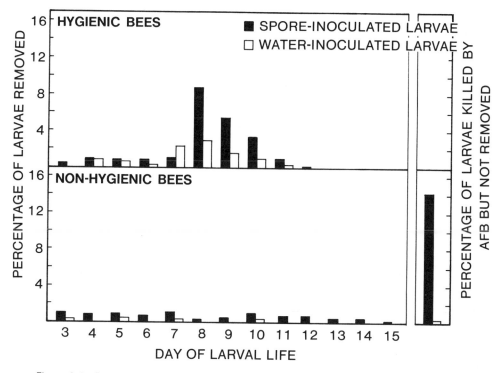

Figure 9.4 Responses of disease-resistant (top) and disease-susceptible (bottom) colonies to larvae inoculated with either American foulbrood spores or water. The disease-resistant colonies skillfully remove all infected larvae before they are 12 days old, and so avoid the disease, but they also remove many healthy larvae. (Modified from Rothenbuhler 1964.)

et al. 1982). The importance of this defense trait was measured by experimentally removing the leafy vegetation around nests to expose them, while leaving other nests unchanged, to serve as controls. Within 7 days of starting the experiment, 4 out of 7 of the exposed nests had been discovered and destroyed by predators, while none of the concealed control nests had suffered any damage (Seeley, unpublished observations). Clearly, the benefits of this single defense trait are enormous, and thus, although no assessment of its costs (difficult flight to and from the nest? poor visibility of the sun for orientation?) has been made, it seems clear that the benefits here exceed the costs.

10 Behavioral Ecology of Tropical Honeybees

The Importance of Studies in the Tropics

The previous seven chapters of this book have summarized the behavioral ecology of one species of honeybee, *Apis mellifera*, as it lives in temperate regions of the world. We have seen that this insect survives in cold climates through many unique adaptations in colony cycle, reproductive process, and techniques of nest building, temperature control, foraging, and colony defense. Throughout this discussion we have been primarily concerned with the refinements in the bees' social behavior which foster survival in a seasonally harsh environment. This view of honeybee behavioral ecology, however, skirts a major portion of the total subject, namely, the ecology and social behavior of honeybees inhabiting the tropics. This chapter is therefore devoted to a discussion of the tropical honeybees, with emphasis on their special significance to ecological studies of insect social behavior.

Although the honeybees in the tropics have traditionally attracted little attention from students of the social bees, their study is essential for a deep, ecological understanding of honeybee social life. One reason for this is that, as was indicated in Chapter 4, the honeybee's social evolution occurred primarily in a tropical environment. Thus, attempts to identify the adaptive origins of the honeybee's social system will often prove most fruitful when pursued with tropical bees. Traits for which this is especially true include reproduction by swarming, regulation of broodnest temperature, and the information-center system of social foraging. A second attraction for expanding honeybee research outside the temperate regions is that certain social behaviors are unique to the tropical forms of honeybees. The most vivid example of this is the mysterious process of colonies performing seasonal, long-distance migrations from forage-poor to forage-rich areas. Although this is a well-developed trait of *Apis dorsata* in southern Asia (Koeniger and Koeniger 1980) and of *Apis mellifera* in central and southern Africa (Fletcher 1978), and surely involves a rich assembly of social mechanisms, our understanding of these colony migrations is almost nil.

Undoubtedly, though, the prime attraction for undertaking studies of the tropical honeybees is that they enable one to conduct analyses of behavioral adaptation using an especially powerful technique—the comparative approach

(Jarman 1982, Clutton-Brock and Harvey 1984). Both intraspecific and inter-specific comparisons will prove rewarding. Among intraspecific studies, prob-ably the richest vein of research is the one involving behavioral and ecological comparisons among temperate and tropical races of *Apis mellifera*. Prelimi-nary efforts in this direction (Winston et al. 1983) indicate that this research will ultimately illustrate how a single insect social system can be adaptively tuned through fine adjustments in colony properties to operate effectively in radically different environments. In general, intraspecific comparisons largely guarantee that any observed differences reflect adaptation to different envi-ronments, rather than profound phylogenetic distinctions which are unrelated to each population's current ecology. Temperate-tropical comparisons be-tween populations of *Apis mellifera* in particular are moreover blessed by the solid baseline for comparison provided by the long history of scientific in-vestigation with the European races of this honeybee.

Interspecific studies with honeybees, like intraspecific ones, provide ma-terial for exploring how one, basic social system can be adapted to function successfully in different ecological settings; but because different species are involved, the morphological, physiological, and behavioral adjustments are more striking. Probably the most revealing interspecific comparative studies with honeybees will be those designed around the three species of *Apis* which live sympatrically across much of southern Asia: *A. florea, A. cerana*, and *A. dorsata*. All three species share such special features of honeybee biology as the dance-language system of communication, nests consisting of vertically hanging combs built of beeswax, barbed stings, (E)–9–oxo–2–decenoic acid as queen substance pheromone, and similarly modified male genitalia, and so are obviously closely related members of one monophyletic group. But despite their phylogenetic proximity, the three species exhibit marked dif-ferences in such ecologically important attributes as worker size (see Fig. 2.3), colony population, nest site (see Fig. 2.4), and worker aggressiveness. Comparing these species should yield insights into how different ecological factors, such as the abundance and distribution of food and predators, have influenced the evolution of particular traits, such as colony size and nest design. Perhaps even more valuable, though, will be the detailed understand-ing that should emerge of how the numerous properties of a social insect colony are tightly interwoven to build effective solutions to such challenges as colony defense, temperature control, and food collection.

African and European Versions of *Apis mellifera*

Although the vast majority of studies on *Apis mellifera* have involved the races native to the temperate climate of Europe, more than two-thirds of the natural distribution area of *A. mellifera* falls within tropical and subtropical Africa. Across this continent this bee inhabits such diverse settings as lowland rain forests, semi-arid savannahs and scrub forests, steamy coastal swamps, and cool mountain ranges. The smallest and the largest, the blackest and the brightest forms of *A. mellifera* exist in Africa. One measure of this diversity, which is a product of adaptation to a wide range of environments, is that systematists of the genus *Apis* recognize no fewer than 11 distinct races of *A. mellifera* across the African continent (Fig. 10.1) (Ruttner 1975b). This discussion will focus on just one of these races, *Apis mellifera scutellata* (formerly called *A. m. adansonii*), the widespread honeybee of the east and south African savannahs. (For simplicity, hereafter this bee will be called the "African bee," while members of the major European races, *A. m. carnica*, *A. m. ligustica*, and *A. m. mellifera*, will be referred to collectively as "European bees.") *Apis mellifera scutellata* has been singled out because it is by far the best studied of the many African races, with significant published research on it conducted both in Africa and South America, the latter work following this bee's introduction into Brazil in 1956 (Michener 1973, 1975).

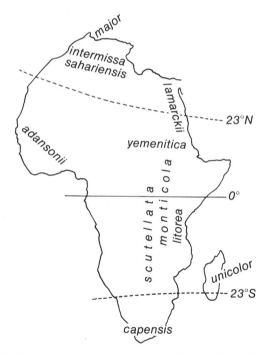

Figure 10.1 Distribution of the races of *Apis mellifera* in Africa. (From Ruttner 1975b and personal communication with F. Ruttner.)

There are limitations to the discussion, however. Reviewing its biology is complicated by the variation in behavior and ecology which occurs across its geographical range within Africa. Moreover, the honeybees in South America, although they are now evidently pure African bees (at least in northern South America) (Daly 1975, Nunamaker and Wilson 1981), must represent only a tiny fraction of the total gene pool of the *scutellata* race. Of the 26 African queens which escaped in Brazil and constituted the nucleus of the population there, all but one, from Tanzania, came from the vicinity of Pretoria, South Africa (Michener 1982). Despite these complications, because the contrasts between African honeybees and their European counterparts are quite bold, we can begin to identify the pivotal differences between the two races that underlie their reciprocal successes and failures in tropical and temperate environments.

This contrast in adaptation between African and European bees is dramatically illustrated by the outcomes of deliberate attempts to transplant these bees between climatic zones. European bees have been successfully introduced throughout temperate regions of the world, including North America, Australia, and northeast Asia, providing large populations of feral colonies as well as productive beekeepers' colonies, but they have always fared poorly in tropical habitats. Over a period of about 40 years, ending in 1965, a massive number of mated European queens were imported into South Africa for introduction into *A. m. scutellata* colonies. Once these colonies lost their African workers, however, and were Europeanized, they began to dwindle (Fletcher 1978). Similarly, European honeybees have never flourished in the South American tropics, especially in the lowlands, despite their repeated introductions over the past 400 years. Beekeepers managed to maintain small apiaries of European bees in the American equatorial zone, but wild colonies in the forests always remained scarce. Thus, before the introduction of African bees, one could collect flower-visiting insects in the tropical forested parts of Brazil, Panama, Venezuela, and neighboring countries for long periods without seeing a honeybee (Michener 1975). Today, some 30 years after the release of African bees in Brazil, they already exist at densities reaching 100 or more colonies per square kilometer, often dominate rich sources of nectar and pollen (Roubik 1978, 1980), and are expanding their range northward by about 500 kilometers annually (Taylor 1977). In utter contrast to this overwhelming biological success in tropical South America, African bees founder in cold temperate regions. Colonies introduced into Poland all perished during their first winter (Woyke 1973b).

It is probably not too gross an oversimplification to state that the critical environmental differences behind these contrasts in adaptive fit are climate and intensity of predation. In cold temperate regions of the world, prolonged periods of subfreezing, winter temperatures alternate with times of summer heat, but in the wooded savannahs of East Africa, the prime habitat of *A. m. scutellata* (Smith 1953), temperatures remain moderate year round, rarely exceeding 32°C or dropping below 10°C, and changing from month to month on average by only 1 to 2°C. The main climatic factor defining the seasons here is rainfall. In the Serengeti ecosystem in northern Tanzania, for example, rains fall mainly from November to May, with the rest of the year quite dry (Schaller 1972, Norton Griffiths et al. 1975). With respect to predation, although no figures are available for predation rates on wild colonies of *A. mellifera* in undisturbed European or African habitats, it seems clear from the extensive honey-hunting by humans in Africa (Guy 1972b, Pager 1973, Nightingale 1976, Crane 1983), and the long lists of honeybee predators cited by African beekeepers (Smith 1960, Guy 1972a, Silberrad 1976, Fletcher 1978), that predation on honeybee colonies has historically been far heavier in Africa than Europe.

What morphological, physiological, or behavioral differences underlie the divergent patterns of adaptation in these two races? Certainly the differentiation does not reflect a gross, qualitative difference in any one attribute, such as thermoregulatory powers or skill in colony defense, but instead exists because of quantitative differences in a broad constellation of traits. What follows is a point by point review of the more important lines of biological tuning identified to date.

Nesting biology. European bees base their selection of a nest site on narrow criteria: a cavity with a volume of 20 to 80 liters, an entrance opening high off the ground, and walls capable of tightly enclosing the bees' combs (see Chapter 6). Such choosiness makes perfect sense for bees whose survival through long, freezing winters depends heavily on a suitably protective nest site. African bees, in contrast, do not face such harsh winters and so can use a wider array of nest sites, a clear advantage wherever unoccupied cavities are scarce. These bees will occupy nearly any cavity larger than 10 to 20 liters which offers some shelter from the weather (Smith 1960, Rinderer et al. 1982). Quite often the protection is minimal, as when the bees build combs beneath tree limbs, in discarded tires, or other essentially open sites.

European and African honeybees also differ markedly in the sizes of their nests. Feral colonies of African bees in Peru evidently cease building comb once it fills a volume of 20 to 25 liters, even if the room available in the nest cavity is far greater (Winston and Taylor 1980). This produces a nest whose comb area covers only 8000 to 11,000 cm^2, far smaller than the 24,000 cm^2 of comb surface found on average in wild colonies of European bees (Seeley and Morse 1976). Smaller nests are superior for African bees, at least in part because these bees do not need to store up a vast quantity of honey for winter survival.

Colonial thermoregulation. Maintaining a warm microclimate inside the nest is critical for honeybee survival through the prolonged periods of subfreezing temperatures in European winters, but is generally not required by honeybees to survive in Africa. This suggests that the techniques of colonial thermoregulation discussed in Chapter 8 for European bees may be less refined in African bees. The facts, though, are not clear. Woyke (1973b) unquestionably found that African bees cannot survive winters in Poland, but this may reflect less a failure to generate heat and conserve warmth than an inability to perform the endocrinological adjustments which enable European bees to live 150 to 250 days in winter as against 30 to 50 days in summer (Sakagami and Fukuda 1968, Fluri et al. 1977, 1982). Equally problematic is the observation that the African honeybee's southward expansion in South America has stalled at 32 to 33°S latitude, in northern Argentina. Though this boundary could reflect a low tolerance of cold, it might also indicate a lack of forage, or the problem just mentioned—failure to adjust physiologically for extended lifespan during a broodless winter period (Taylor 1977, Taylor and Levin 1978). *Apis mellifera scutellata* does experience seasonal periods of subzero temperatures in certain regions of its distribution, such as the Drakensberg mountains in South Africa, so one should expect that it can perform thermally efficient clustering and survive at least brief periods of freezing temperatures (Fletcher 1978). Yet when Núñez (1979b) conducted direct comparisons of African and European bees in a controlled environment, the African bees proved less skilled at forming tight, well-insulated clusters. Groups of 200 workers were placed with brood and food in a 10°C refrigerator, and although both European and African bees kept the brood warm at 34°C, the European bees quickly assembled into a motionless cluster while the Africans ran about continuously in "febrile agitation." Clearly, better documentation is needed of the African honeybee's powers of thermoregulation.

Food collection and storage. Efficient foraging and massive hoarding of honey are key elements in the European honeybee's strategy for survival through long, cold winters (see Chapters 4 and 7). One might expect a priori that African honeybees possess a weaker tendency to amass food reserves since they do not face freezing, flowerless winters and would probably benefit from converting forage rapidly into bees for colony growth and reproduction. However, the observations on this matter are frustratingly contradictory. On the one hand, direct comparisons of African and European bees in the laboratory, where the volume of sugar syrup stored in 7 days by 30 bees was measured, indicate a slightly greater hoarding tendency in European bees, though the difference was only about 10 percent (Rinderer et al. 1982). Also, when the honey stores of wild colonies are compared for the two types of honeybee, a threefold difference is found, with European colonies leading the African ones (Seeley and Morse 1976, Fletcher 1978, Winston et al. 1981). Furthermore, the observation that African bees often cope with local forage dearths by migrating to richer areas, rather than remaining in one location and falling back on food reserves (see below), suggests that they rely less on food storage than do European bees. On the other hand, beekeepers in South Africa (Fletcher 1978) and in South America (Michener 1975) report excellent annual honey crops—50 to 200 kilograms per colony—with African bees. Perhaps, though, this reflects certain apicultural manipulations of the bees, such as supplying colonies with a superabundance of empty combs, and so has little relevance to the way African bees live in nature.

Both African and European bees are expected to be efficient foragers in their respective environments, since a honeybee colony's survival and reproductive powers stem ultimately from its ability to extract resources from the environment. There is some evidence of a divergence in foraging techniques between these tropical and temperate region bees, presumably through adaptation to different foraging conditions. Direct comparisons by South American beekeepers of the honey yields of the two types of bees showed marked superiority of the African bees in a tropical setting. Portugal-Araújo (1971), for example, reports average honey productions of 8.8, 19.2, and 35.5 kilograms per hive for 10 colonies each of brown European (*Apis mellifera mellifera*), yellow European (*A. m. ligustica*), and African (*A. m. scutellata*) bees, respectively, during a nectar flow in Brazil. Presumably the colonies were matched in size and location. Exactly how African bees excel as foragers in a tropical environment remains a mystery. It has been suggested that many tropical plants secrete nectar primarily during the cool hours of dusk and darkness, and that African bees can better exploit these sources because they fly at lower temperatures and lower light levels than European bees (Smith 1953, 1958, Fletcher 1978). Also, African bees appear more responsive to changes in forage patch quality (Núñez 1979a) and initiate foraging at an earlier age than European bees (Winston and Katz 1982), but the precise connection between these behavior patterns and differences in foraging effectiveness in the tropics remains obscure.

Colony mobility. The most intriguing distinction between African and European bees lies in the relative mobility of their colonies. European bees rarely abandon nests, undoubtedly because of the immense risk in doing so in temperate regions, where the time available to establish a new home is so short. In marked contrast, African bees frequently abscond, with colonies completely vacating one nest site to reside elsewhere (Guy 1976, Silberrad 1976, Winston et al. 1979). Clearly, these bees face far more flexible time and energy constraints in shifting nest sites and so have been free to explore the benefits of colony migration. What is especially fascinating about absconding by African bees is that it evidently takes two forms: short-distance (generally less than 10 kilometers) and long-distance (up to 100 kilometers or more) moves. The reasons behind the relatively short moves are numerous but fairly obvious. They include heavy attacks by predators, a colony outgrowing a small nest cavity, and excessive exposure to sun or rain (Chandler 1976, Fletcher 1978). The long-distance moves evidently serve a different purpose, namely, movement from a resource-poor to a resource-rich environment. Well-documented examples include the migration of thousands of colonies into the Tugela valley of Natal and Zululand, South Africa, when the vast expanse there of *Isoglossa eckloniana* shrub comes into bloom (Fletcher 1978), and the seasonal alternation of bees between the mountains and the Rift valley in Kenya (Nightingale 1976).

Although these moves are consistently associated with a dearth in forage (Chandler 1976, Winston et al. 1979), one would like to know more about the selective forces favoring colony migration. In particular, do colonies undertake their journeys as a last resort, to avoid outright starvation, or do they make these moves to improve on an adequate, but inferior, food supply? Winston, Otis, and Taylor (1979) observed colonies of African bees abscond when forage was scarce but the bees' nests still contained considerable honey and pollen, thus indicating that colonies will abscond without the threat of imminent starvation. Whatever the circumstances surrounding migration, this behavior depends on spatial heterogeneity in resources over the geographic range which colonies can travel. Heterogeneity in rainfall, and thus probably also in flowering and bee forage, is well documented in the Serengeti–Mara ecosystem of Tanzania and Kenya, where it has been studied in connection with the migrations of wildebeest (Pennycuick 1979). Here, and probably throughout much of the range of African bees, spatial differences in rainfall have both predictable (between woodlands and plains) and unpredictable (caused by local rainstorms) components, the net product of which is marked differences in plant growth (and foraging conditions for bees?) between locations spaced 20 to 100 kilometers apart.

The orientation mechanisms of migrating honeybee swarms remain an almost total mystery. Are the bees' travels based on genetically programmed movements in certain directions in particular seasons, long-distance perception of wind-borne floral odors, scout bees surveying the foraging opportunities within an immense area, or still some other technique? Knowing the weight of an unengorged African bee (60 mg, Otis 1982a), her weight-specific metabolic rate when in flight (110 mg sugar $\cdot$ g bee^{-1} $\cdot$ hr^{-1}, Heinrich 1980), her flight speed (24 km/hr, von Frisch 1967), and her maximum fuel load (about 60 mg, Otis et al. 1981), one can calculate that a scout bee can fly out and back about 55 kilometers (110 km total) on one load of honey. Thus it is at least conceivable that direct monitoring of foraging opportunities across great distances plays a role in orienting a colony's long-distance travels. Similar calculations based on measurements of the sugar in the crops of absconding bees indicate that swarms carry sufficient fuel to fly about 90 kilometers (Otis et al. 1981), probably a sufficient range to carry them to better forage in one move.

The preparations for migrations are remarkably orderly. Figure 10.2 shows that the process begins with a gradual decline in brood rearing starting some 20 to 25 days before departure. The bees cease rearing larvae 10 to 15 days before leaving, evidently eating any larvae encountered after this time. Also, by this point the queen's egg-laying has fallen to about one-fourth of its starting rate. Over the final two weeks the bees steadily consume their pollen stores, and the queen's egg-laying continues to fade, so that finally, when all the sealed brood has emerged and the swarm flies off, few eggs and little pollen remain behind.

Defense behavior. When a biologist familiar with European bees first encounters a colony of African bees, he will probably be astonished by the defensive ferocity of these tropical bees. Whereas prying the lid off a hive housing a strong colony of European bees will generally elicit only a dozen or so stings, the same act with African bees is likely to result in a hundred or more stings. Stort (1974, 1975) measured this difference by dangling a 2-cm-diameter black leather ball at the entrance of a hive, jerking it up and down for 60 seconds, and counting the stings buried in the leather ball and in his gloves. Trials with African bee colonies produced 61 and 49 stings, on average, in the ball and gloves, while the same figures for tests with

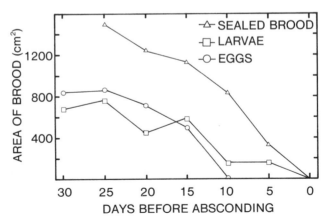

Figure 10.2 Colonies shut down brood rearing in preparation for migration to an area richer in forage. (Modified from Winston et al. 1979.)

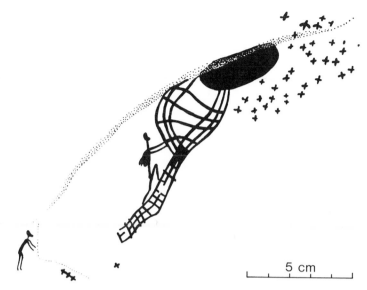

5 cm

Figure 10.3   Honey hunter climbing a ladder to reach a nest of honeybees. Rock paint-ing (date unknown) in Eland Cave, Drakensberg Mountains, Natal, South Africa. (From Pager 1971.)

European bees were only 26 and 0 stings. Tests with a similar design conducted by Collins et al. (1982), which featured tighter control of colony size, revealed an even higher, nearly sixfold, ratio of sting deliveries between paired African and European colonies.

Undoubtedly the African honeybee's stronger sting defense traces ulti-mately to a history of greater predation by vertebrates, and there can be little question that the primate *Homo sapiens* has been and continues to be the African bee's foremost vertebrate foe. Honey-hunting by humans occurs throughout the range of *Apis mellifera scutellata*. Furthermore, the numerous rock paintings of southern Africa, frequently depicting bees, honey combs, and human honey gatherers, provide graphic testimony that man's history as a predator on honeybees spans at least two thousand years (Pager 1973, Crane 1983) (Fig. 10.3). Present-day descriptions of the honey-hunting and bee-keeping techniques of the east and south African peoples provide a detailed picture of the defense problem which these bees have faced (Smith 1958, Brokensha et al. 1972, Guy 1972b, Kigatiira 1984). Sometimes this predation involves simply locating a thriving colony, then returning under cover of darkness and smoke to destroy the colony and steal its honey-filled combs. More often, hives built of bark or hollowed-out logs are wedged in trees as homes for wild swarms, and periodically the hives' owners harvest honey from the occupied hives, usually without killing the colonies. With both methods, though, colonies suffer heavy damage and those able to mount rapid, massive counterattacks by stinging guard bees surely have been favored by natural selection.

Numerous nonhuman predators also threaten honeybees in Africa, particularly birds such as the alpine swift (*Apus melba*) and the bee-eater (*Merops apiaster*), and wasps such as *Palarus latifrons*. However, these three enemies catch bees in flight, generally when they are well away from their nest. Perhaps the only other animal besides *Homo* whose predatory behavior has molded the African bee's explosive defense response is the honey badger or ratel (*Mellivora capensis*). Like man, the honey badger is strong enough to tear open nests, and though it lacks defense by smoke, its astonishingly tough skin helps repel stings (Walker et al. 1975). The destructive powers of *Mellivora capensis* toward honeybees are probably further strengthened by its mutualistic relationship with a certain species of bird, appropriately named the greater honey-guide (*Indicator indicator*). Queeny (1952) and Friedmann (1955) describe in detail how this bird will lead honey badgers (or humans) to bee nests, starting by approaching its collaborator and performing a piercing series of churring notes to attract its attention. Once this has been gained, the bird guides the honey badger step-by-step over a distance of up to several hundred meters to a nest of bees. At this point the guide waits quietly for the other animal to rip out the combs and take its fill. The bird's reward for its labor is the beeswax wreckage of the nest, which it digests for energy.

Investment in reproduction. Here we deal with the most fundamental line of divergence between African and European bees: ratio of investment in colony survival and reproduction. African bees invest several times more heavily in reproduction and less intensively in survival as compared to European bees. The clearest expression of this difference is the number of swarms produced annually by a colony: 6–12 for African bees (Winston et al. 1981, Otis 1982b), but only 2–3 for European bees (Winston 1980). (The swarms of the two types of bees are closely matched in size.) A second measure is the quantity of honey present in colonies following acts of swarming: 920 ± 160 and 2800 ± 460 cm^2 of honey-filled combs in African and European nests, respectively (Winston et al. 1981). The resources represented by this honey could have been converted into bees, and thus devoted to swarming, but are instead set aside as an investment in the parent colony's future survival.

The ultimate origins of these differences probably lie in the climate and predation experienced by African and European bees in their native environments. As we have seen (Chapter 4), colonies of European bees have only about 2 months each year when they can swarm, given the brevity of temperate zone summers and the considerable time required by new colonies to assemble their large food supplies for winter. African bees live in a milder climate and hence experience less restrictive time constraints on swarming. In Zambia, for example, there are two 2- to 3-month swarming seasons each year (Silberrad 1976), and in the coastal savannahs of French Guiana (South America), African bees are free to swarm throughout an 8-month period each year (Winston et al. 1979). Since both African and European colonies require 50 to 70 days to rebuild to swarming strength following a bout of swarming, these swarming statistics imply that colonies of African bees can cycle through two to four bouts of swarming annually, while over the same time period European colonies squeeze in just one or occasionally two episodes of swarming (Winston et al. 1981, Otis 1982b). Because both types of bee produce 2 to 3 swarms per swarm cycle, it is clear that the difference in reproductive investment between African and European bees has been strongly shaped by the climatic differences which give these bees such different temporal opportunities for swarming.

While a milder climate provides African bees with a greater opportunity for investing in reproduction, high levels of colony destruction by predators were probably critical in making heavy investment in reproduction, rather than survival, advantageous. Only in an environment kept "empty" by high levels of colony mortality will natural selection steadily favor the high reproductive rate which characterizes African bees.

Asian *Apis*: Contrasts in Adaptation among Closely Related Species

The three honeybee species which live in southern Asia—*Apis florea, A. cerana*, and *A. dorsata*—present a rich puzzle to the biologist interested in the adaptive modification of insect social behavior. On the one hand, they are closely related phylogenetically, and as true honeybees, they share such special attributes as the dance language and vertical combs built of pure beeswax. But on the other hand, the three species exhibit numerous contrasts in behavior and morphology, a sampling of which are listed in Table 10.1 and illustrated in Figures 2.3 and 2.4. What is the adaptive significance of these interspecific differences? Why, for example, does *A. dorsata* possess workers which are 5 times heavier, colonies 30 times more massive, and colonial foraging areas 100 times broader than those of *A. florea*?

Such questions can be attacked from two opposite directions. One, a trait-oriented approach, starts with selecting a particular feature of the organisms under study, such as colony size or nest site in the case of honeybees, and then working to identify the forces of natural selection underlying the interspecific differences in this attribute. This method possesses several attractions,

Table 10.1

Comparison of the three honeybee species of southern Asia. Based on Dhaliwal and Sharma 1974, Koeniger and Vorwohl 1979, Koeniger and Koeniger 1980, Seeley et al. 1982, and references cited therein.

	A. florea	A. cerana	A. dorsata
Worker body size (mg, partially loaded)	32	54	155
Nest site	Branch of shrub	Cavity	Tree limb or cliff
height (m)	Low (< 5)	Low (< 2)	High (> 15)
visibility	Hidden	Conspicuous	Conspicuous
dispersion	Widely spaced	Widely spaced	Clustered
Colony population	6000	7000	37,000
aggressiveness	Low	Low	High
movements	Local	Stable	Migratory
foraging area (km²)	Small (<3?)	Small (< 10?)	Large (> 300?)
mass (kg)	0.2	0.4	6.0

especially the high power of analysis which comes from focusing on a specific trait, and the possibility that the work will uncover a general pattern in biological adaptation for the trait under investigation. The second approach starts with the other half of the process of adaptation, focusing on a particular ecological problem, such as predation, thermoregulation, or food collection, and then seeking to understand each species' distinctive solution to this problem. The appeal of this method is its effectiveness in revealing the mutually adaptive fit of an organism's many properties, and thus showing how complex patterns of differences between species can represent simpler, more fundamental, adaptive themes. Whichever approach one begins with, ultimately there is convergence with the other style of analysis. The adaptive design of any one trait can rarely be fully understood by viewing the trait in isolation. Adaptation involves the whole organism, and in the case of social insects, often the whole colony. Thus one must eventually adopt a broad view of each species' patterns of adaptation. Reciprocally, if one begins with an overview of a species' ecological problems, and so first assembles a sense of its basic design themes, one must eventually narrow the analysis down to a few key traits. Only through close examination of the details of species differences in behavior, morphology, and physiology can one truly appreciate how different patterns of adaptation take shape.

A trait-oriented analysis. Because size constrains virtually all aspects of an organism's interactions with the environment, the differences in worker size among the Asian honeybees represent perhaps their most fundamental line of divergence. Thus, understanding the adaptive significance of their body-size differences may be the key to understanding their adaptive radiation. As a first step toward explaining why the workers of the Asian *Apis* differ so markedly in size, we shall try to identify which ecological contrasts are tightly correlated with this morphological difference. Then we shall discuss how future work can help distinguish which of these ecological relationships caused the size divergence and which ones evolved only secondarily, once the size differentiation had unfolded.

A bee's size profoundly influences the way it can counter the threat of predators. In general, the larger their workers, the harder it is for colonies to conceal themselves from visually hunting predators, but the easier it is for them to defend themselves once discovered. Thus when walking through the forests of Thailand, I find it easy to spy *Apis dorsata* nests hanging in tall trees, or *A. cerana* workers darting in and out of their nests' entrances, but can just as easily pass right by an *A. florea* colony, missing completely the aerial traffic of its tiny foragers. But once I find an *A. florea* colony, collecting it poses no danger. Workers of *A. florea* possess such small stings that they often have difficulty implanting them in human skin (and presumably the skin of other attackers). *Apis dorsata* colonies, in contrast, are best approached with high caution. Their stings have no trouble penetrating both one's clothing and skin simultaneously (Fig. 10.4). *Apis cerana* is a competent stinger as well. Size also affects the bees' defensive powers beyond their stinging ca-

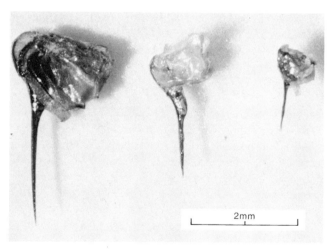

Figure 10.4 Size comparisons of the worker bee stings of Asian honeybees. From left to right: *Apis dorsata, A. cerana, A. florea.*

pabilities. *Apis florea* workers are no match in direct combat against weaver ants (*Oecophylla smaragdina*), a major predator on honeybees, and so they rely upon barriers of sticky plant resins deposited on the branches supporting their nests for defense against these ants. Chemical weapons against ants are lacking in *A. cerana* and *A. dorsata*, whose larger workers quickly seize and tear apart any intruding weaver ants.

A second ecological difference which correlates closely with the differences in worker size is an individual bee's thermal niche. The smaller the bee, the higher its surface-to-volume ratio, and the less its body temperature rises above the ambient temperature when the bee is active. Thus the temperature range over which small bees are active may be shifted upward relative to that of larger bees. Unfortunately, the thermoregulatory powers of the Asian bees have not been compared empirically, so this discussion will be limited to exploring the matter by considering the allometric relationships among body size, metabolic rate, and heat conductance. The critical idea here is that metabolic rate and heat conductance scale to body size in a way which is described by the equation:

$$Y = a \cdot X^b, \tag{10.1}$$

where X = body mass, Y = metabolic rate or conductance, and a and b are fitted constants (Peters 1983). For heterothermic moths, and birds and bats, and so probably also heterothermic bees such as honeybees, the rate of energy metabolism during flight scales with mass to approximately the 0.75 power ($b = 0.75$) (Bartholomew 1981). The metabolic rate of workers of *Apis mellifera*, whose mass is approximately 0.000100 kg each, has been measured to be about 503 W/kg, or 0.0503 W/bee when in flight (Heinrich 1980). These data allow us to predict that honeybees of different masses (X, in kg/bee) will have metabolic rates in flight (R, in W/bee) described approximately by the following equation:

$$R = 50.3 \ X^{0.75}. \tag{10.2}$$

To predict a bee's body temperature, we must also find a formula describing heat conductance as a function of body mass, since a bee's temperature is a dynamic equilibrium between heat gain and heat loss. An animal's conductance is the product of three quantities: its cooling constant (which scales allometrically with body size), the specific heat of its tissues, and its body mass. In the absence of forced convection (i.e., when not in flight), all bees, flies, dragonflies, and moths that have been examined show a similar relationship between cooling constant (λ, in sec^{-1}) and body mass (kg/individual) (May 1976):

$$\lambda = 0.00027 \ X^{-0.38}. \tag{10.3}$$

Given the generally accepted value for the specific heat of animal tissue of $3400 \; \text{J} \cdot \text{kg}^{-1} \cdot {}^{\circ}\text{C}^{-1}$, we can now write a formula for conductance (C, in $\text{W} \cdot {}^{\circ}\text{C}^{-1} \cdot \text{bee}^{-1}$) as a function of body mass:

$$C = 0.00027 \; X^{-0.38} \; (\sec^{-1}) \cdot 3400 \; (\text{J} \cdot \text{kg}^{-1} \cdot {}^{\circ}\text{C}^{-1}) \cdot X \; (\text{kg/bee})$$
$$= 0.92 \; X^{0.62}. \tag{10.4}$$

We can now estimate the air-body temperature difference at which workers of each species experience a stable body temperature when in flight and without active thermoregulation. This occurs when the ratio P of metabolic rate to heat flux out of a bee equals 1. The rate of heat loss from a bee is the product of its conductance and the temperature gradient between the bee's body and the ambient environment. Thus:

$$P = \frac{R}{C \cdot (T_B - T_A)} = \frac{50.3 X^{0.75}}{0.92 X^{0.62}(T_B - T_A)} = \frac{54.7 X^{0.13}}{(T_B - T_A)}, \tag{10.5}$$

where T_B and T_A denote body and ambient temperatures.

The predictions of this equation for bees with the body masses of the three Asian honeybees' workers are shown in Figure 10.5. This plot shows that the workers of these three species are expected to be at thermal equilibrium in temperature gradients ($T_B - T_A$) of 14.5, 15.5, and 17.5°C. These numbers are most applicable to bees shivering while standing still since the conductance formula used is for bees in still air. For bees in flight, where heat flux is accelerated by forced convection, the conductance values will be higher and the temperature gradient at which each species is expected to be at thermal equilibrium will be correspondingly smaller. It is also likely that the differences between species are underestimated in Figure 10.5 because the larger the honeybee, the thicker its thoracic coating of insulating hair (personal observations), and differences in degree of insulation have a much greater effect on heat transfer in moving air than in still air (Church 1960)—the situation represented by the Figure 10.5 curves. Clearly there is a need for both laboratory and field studies on thermoregulation by the Asian honeybees. To date, there are no field studies explicitly comparing the abilities of these bees to work a standard food source at high and low temperatures, though Murrell and Nash (1981) do report that *A. cerana* workers consistently arrived each morning one hour before *A. florea* workers at a field of toria (*Brassica campestris* var. *toria*) in Bangladesh during the relatively cool months of December and January.

Besides differences in defense and thermoregulation opportunities, the three species of Asian honeybee undoubtedly experience marked disparities in the foraging process associated with their differences in worker size. To begin

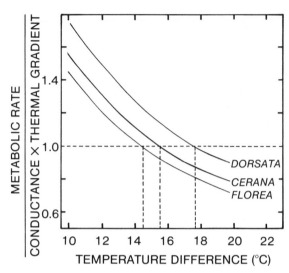

Figure 10.5 Ratio of heat production to heat loss for workers of the Asian honeybees as a function of the temperature gradient between bee and environment. When the ratio equals one, the bee is in thermal equilibrium.

to understand these differences, it is useful to estimate the relative flight speeds, relative individual transport costs, and relative mass-specific transport costs of the three species. The appropriate scaling coefficients for these al-lometric relationships are given in Table 10.2. These calculations predict some major differences in the foraging energetics of these bees. For example, an *Apis dorsata* worker in flight would consume more than three times as much energy as an *A. florea* worker to cover a given distance, but could carry a given mass of forage across a given distance with 30 percent less energy. Such relationships suggest that the larger bees would require richer floral rewards to break even energetically when foraging, but that given a food source which is profitable to all three species, the larger bees could retrieve food from it more cheaply across a given distance, or could carry it back over greater distances for the same cost than could the smaller bees. Although the controlled comparisons which would empirically test these ideas are lack-ing, various observations match with the predicted differences in foraging behavior. One is that *A. florea* foragers will continue foraging at a patch of flowers long after *A. cerana* foragers have abandoned it (Murrell and Nash 1981), perhaps because of the smaller bee's lower minimum floral reward for profitability. Another is the correlation between worker size and foraging range among the three species (Lindauer 1956), although the differences in

Table 10.2

Estimations of parameters influencing the foraging energetics of the honeybees of southern Asia. The values shown for each parameter are calculated relative to the value expected for a bee with the mass of an *A. florea* worker. Each parameter, except body mass, was assumed to scale with body mass to the power (value of *b*) indicated.

Species	Relative body mass	Relative flight speed ($b = 0.14$)[a]	Relative mass-specific transport cost ($b = -0.23$)[b] ($J \cdot kg^{-1} \cdot m^{-1}$)	Relative transport cost ($b = 0.77$)[b] ($J \cdot m^{-1} \cdot bee^{-1}$)
A. florea	1.0	1.00	1.00	1.00
A. cerana	1.7	1.07	0.89	1.50
A. dorsata	4.6	1.24	0.70	3.24

[a] Scaling coefficient from Bonner 1965.
[b] Scaling coefficient from Tucker 1970.

foraging range (see Table 10.1) appear far greater than are predicted simply from consideration of mass-specific transport costs. For example, the maximum flight range of *A. dorsata* workers (about 10 kilometers) appears to be some 10 times greater than that of *A. florea* workers (about 1 kilometer), yet the mass-specific transport cost for the former bee is predicted to be only about one third lower than that of the latter bee.

The preceding discussion is little more than an introduction to the ecological implications of worker size for the Asian honeybees, but it does hint at the richness of the subject. At the same time, this analysis points out a fundamental problem in trait-oriented studies of adaptation: the problem of identifying which ecological feature in an array of ecological differences is responsible for the difference in behavior, morphology, or physiology under investigation. At present, one cannot draw any firm conclusions concerning the adaptive significance of the differences in worker size among Asian honeybees. A common assumption made by evolutionary ecologists is that the feeding niche occupied by a species constrains evolutionary change due to other forces (Clutton-Brock 1977), but this seems questionable for these bees. Predation seems at least as important as foraging in their ecology. I am optimistic that future fieldwork will help solve this particular mystery and others like it. Field studies can, for example, directly examine the ecological consequences of body size through controlled species comparisons in such matters as ability to repel certain predators, ability to fly at low and high temperatures, and willingness to forage at food sources differing in their quality and distance from the nest. It is entirely possible that such real-world studies will indicate that some of the suggested ecological constraints imposed by body size are actually negligible, possibly because there exist behavioral or physiological adjustments which counterbalance certain size effects.

A problem-oriented analysis. The second approach to identifying the adaptive significance of behavioral differences between closely related species is illustrated by a comparative study of the Asian honeybees' strategies of colony defense (Seeley et al. 1982). Here the focus is on one ecological problem—predation—and the interspecific differences associated with the three species' distinct solutions to this one problem.

Defense is a logical theme of study in ecological investigations of social bees. Because their nests are filled with nutritious brood and energy-rich honey, colonies of social bees suffer frequent attacks by hungry outsiders. In Khao Yai National Park in northeast Thailand, for example, *Apis florea* and *A. cerana* colonies face probabilities of nest destruction of about 0.25 and 0.10 per month during the dry season. The foes which create these statistics are a diverse lot and include weaver ants (*Oecophylla smaragdina*), giant social wasps (*Vespa* spp.), tree shrews (*Tupaia glis*), agamid lizards, the Eurasian honey-buzzard (*Pernis apivorus*), the Malayan honey bear (*Helarctos malayanus*), and various primates, such as rhesus monkeys (*Macaca mulatta*) and humans (Seeley et al. 1982). The logic of singling out predation as a key selective force on the Asian honeybees gains further support from the bees' well-documented differences in worker size and nest site (cavity versus open air). As a rule, size greatly affects an animal's fighting ability, and animals which nest in cavities generally experience a radically different defense situation from those nesting in the open (Lack 1968). As we shall see, this reasoning was borne out by the fieldwork. The three species of Asian *Apis* possess markedly different strategies of colony defense, and the behavioral and morphological elements forming these strategies extend throughout each species' biology.

Defense by *Apis florea* colonies against visually hunting predators relies primarily on avoiding detection. To this end, these bees nest low in dense, shrubby vegetation, build small, widely dispersed nests, and will move to a new nest site if their current one loses its cover, as frequently occurs during the dry season when many plants drop their leaves. To minimize the nest construction costs associated with these frequent moves, these bees have the habit of salvaging wax from the old nest and incorporating it in their new home. Their nest concealment is not perfect, however, and if detected, these bees will attempt to repel any intruder. Roughly 80 percent of a colony's workers serve as guards, forming a multi-layered protective blanket which covers the colony's single comb (Fig. 10.6). When an intruder approaches the nest, a salvo of guards, up to several hundred individuals, will launch

Figure 10.6   Nest of *Apis florea* with (left) and without (right) the protective curtain of bees.

nearly synchronously from the protective curtain out at the enemy and attempt to sting it. Unfortunately for the bees, their small size renders such counterattacks fairly ineffective and most attacks by *Vespa* wasps and vertebrate predators end with the bees abandoning their nest. An entirely different line of adaptation provides defense against the deadly weaver ants. These are large ants, fully as large as *A. florea* workers in overall body length, and they have no trouble locating the honeybees' nests. However, by nesting on slender branches, and coating these branches with sticky plant resins, colonies of *A. florea* effectively block invasions by these ants.

Apis dorsata resembles *A. florea* in nesting in the open, but its system of colony defense has evolved along a separate pathway. This large bee does not try at all to hide its nests, but instead positions them high in the crowns of the tallest forest trees or beneath rock overhangs, locations which leave the nests conspicuous but fairly safe, far from terrestrial predators. Its defense against those predators which can reach the lofty nests, such as birds and lizards, consists of explosive defense attacks. These far outscale the counterattacks of *A. florea* colonies, in part because *A. dorsata* colonies possess some six times as many members on average, but mainly because their guards' stings are far more effective in penetrating fur, feathers, and skin.

Apis cerana colonies employ yet a third, highly distinctive defense strategy, one which is built around this bee's habit of nesting inside cavities. Unlike *A. florea*, it seems to make no special attempt to hide its nests. Even though the nests proper are tucked out of sight, they are readily located by the conspicuous flight of these relatively large bees passing in and out of low, clearly visible entrance holes. And unlike *A. dorsata*, these bees are not fiercely aggressive. When under attack, most of a colony's guards retreat inside the protective nest cavity. Even more telling is the way beekeepers can open hives of these bees with little fear of being strongly stung. Defense by *A. cerana* colonies instead centers upon finding a nest cavity with strong walls and a small entrance opening, one usually less than 30 cm² in area. Large predators thus usually cannot gain access to this bee's nest, and small intruders, such as ants, are repelled by a phalanx of guards stationed in the entrance tunnel. Because the nests of *A. cerana* colonies receive automatic protection from their surrounding walls, these bees can afford to invest less in defense. One indication of this is their low aggressiveness. Another is the thin cover of bees over the combs, rarely more than one layer thick, and thus quite unlike the three to six layers which typically form the protective curtains of *A. florea* and *A. dorsata* colonies. Several further traits associated with occupying protective cavities include relatively small colonies, since mass defensive attacks are not needed; nests of multiple combs, to fit the needed comb area inside a cavity; and well-developed techniques of nest ventilation.

The aim of presenting these capsule summaries of each species' defense system has been to demonstrate the wealth of interspecific differences associated with the three species' distinct solutions to the problem of predation. The catalogue of relevant properties includes worker size, nest height and visibility, comb area, number of combs, colony population size, labor allocation to defense, colony mobility, and colony aggressiveness. In short, each species' defense strategy consists of a broad web of numerous, tightly interwoven lines of adaptation.

But does this fact provide a full answer to the puzzle of why the species differ so markedly? Evidently not. For one thing, there remain differences which appear unrelated to colony defense, such as the performance of long-distance migrations by *Apis dorsata* but not by *A. florea* or *A. cerana* colonies. For another, and more fundamentally, some of the interspecific differences which seem to contribute to each species' special defense strategy may really trace to other selective pressures. For example, with respect to *A. florea* and *A. dorsata* colonies, does the high percentage of a colony's bees devoted to forming a protective curtain reflect selection for improved defense or for superior nest temperature control? Similarly, do the species' differences in colony size and worker size trace to different strategies of defense, or foraging, or still some other function? Ideally, better understanding of the fine details of each species' biology, together with certain manipulative studies, will resolve these sorts of ambiguities. For instance, by experimentally varying the thickness of a colony's protective curtain and monitoring its thermoregulatory powers, one could determine the importance of curtain thickness to nest temperature control. Also, a better understanding of the design of each species' foraging system, and especially the importance of worker size and colony size in these foraging strategies, is likely to clarify the energetic constraints produced by these two properties.

At present it is unclear how well we will ever identify the causal relationships between ecology and colony design for the Asian honeybees. Perhaps we will ultimately be limited to identifying clusters of functionally related traits and will never really know what propelled the adaptive differentiation of these bee societies. What seems more likely, however, and what I prefer to believe, is that the honeybee's remarkable openness to scientific investigation, when further exploited in comparative studies, will enable us to pinpoint the ecological origins of these three species' marked differences in morphology and social behavior.

Literature Cited

Adam, B. 1968. *In Search of the Best Strains of Bees*. Walmar Verlag, Zell Weierbach, West Germany.

Adams, J., E. D. Rothman, W. E. Kerr, and Z. L. Paulino. 1977. Estimation of the number of sex alleles and queen matings from diploid male frequencies in a population of *Apis mellifera*. *Genetics* 86:583-596.

Alexander, R. D. 1974. The evolution of social behavior. *Annual Review of Ecology and Systematics* 5:325-383.

Alfonsus, E. O. 1933. Zum Pollenverbrauch des Bienenvolkes. *Archiv für Bienenkunde* 14:220-223.

Alford, D. V. 1975. *Bumblebees*. Davis-Poynter, London.

Allen, M. D. 1956. The behaviour of honeybees preparing to swarm. *British Journal of Animal Behaviour* 4:14-22.

Allen, M. D. 1958. Drone brood in honey bee colonies. *Journal of Economic Entomology* 51:46-48.

Allen, M. D. 1959a. Respiration rates of worker honeybees at different ages and temperatures. *Journal of Experimental Biology* 36:92-101.

Allen, M. D. 1959b. Respiration rates of larvae of drone and worker honey bees, *Apis mellifera*. *Journal of Economic Entomology* 52:399-402.

Allen, M. D. 1960. The honeybee queen and her attendants. *Animal Behaviour* 8:201-208.

Allen, M. D. 1963. Drone production in honey-bee colonies (*Apis mellifera* L.). *Nature* 199:789-790.

Allen, M. D. 1965a. The production of queen cups and queen cells in relation to the general development of honeybee colonies and its connection with swarming and supersedure. *Journal of Apicultural Research* 4:121-141.

Allen, M. D. 1965b. The effect of a plentiful supply of drone comb on colonies of honeybees. *Journal of Apicultural Research* 4:109-119.

Allen, M. D. and E. P. Jeffree. 1956. The influence of stored pollen and of colony size on the brood rearing of honeybees. *Annals of Applied Biology* 44:649-656.

Anderson, E. J. 1948. Hive humidity and its effect upon wintering of bees. *Journal of Economic Entomology* 41:608-615.

Armbruster, L. 1921. Vergleichende Eichungsversuche auf Bienen und Wespen. *Archiv für Bienenkunde* 3:219-230.

Avitabile, A. 1978. Brood rearing in honey bee colonies from late autumn to early spring. *Journal of Apicultural Research* 17:69-73.

Avitabile, A., R. A. Morse, and R. Boch. 1975. Swarming honey bees guided by pheromones. *Annals of the Entomological Society of America* 68:1079-1082.

Bailey, L. 1952. The action of the proventriculus of the worker honeybee. *Journal of Experimental Biology* 29:310-327.

Bailey, L. 1967. The effect of temperature on the pathogenecity of the fungus, *Ascosphaera apis*, for larvae of the honey bee, *Apis mellifera*. In: *Insect Pathology and Microbial Control*. P. A. van der Laan, ed., pp. 162-167. North Holland, Amsterdam.

Bailey, L. 1981. *Honey Bee Pathology*. Academic Press, London.

Bamrick, J. F. 1967. Resistance to American foulbrood in honeybees. VI. Spore germination in larvae of different ages. *Journal of Invertebrate Pathology* 9:30-34.

Bamrick, J. F. and W. C. Rothenbuhler. 1961. Resistance to American foulbrood in honey bees. IV. The relationship between larval age at inoculation and mortality in a resistant and in a susceptible line. *Journal of Insect Pathology* 3:381-390.

Bartholomew, G. A. 1981. A matter of size: an examination of endothermy in insects and terrestrial vertebrates. In: *Insect Thermoregulation*. B. Heinrich, ed., pp. 45-78. Wiley, New York.

Bastian, J. and H. Esch. 1970. The nervous control of the indirect flight muscles of the honey bee. *Zeitschrift für Vergleichende Physiologie* 67:307-324.

Beetsma, J. 1979. The process of queen-worker differentiation in the honeybee. *Bee World* 60:24-39.

Betts, A. D. 1920. The constancy of the pollen-collecting bee. *Bee World* 2:10-11.

Betts, A. D. 1935. The constancy of the pollen-collecting bee. *Bee World* 16:111-113.

Beutler, R. 1953. Nectar. *Bee World* 34:106-116, 128-136, 156-162.

Beutler, R. 1954. Über die Flugweite der Bienen. *Zeitschrift für Vergleichende Physiologie* 36:266-298.

Blum, M. S., H. M. Fales, K. W. Tucker, and A. M. Collins. 1978. Chemistry of the sting apparatus of the worker honeybee. *Journal of Apicultural Research* 17:218-221.

Blum, M. S., H. F. Novak, and S. Taber. 1959. 10-hydroxy-Δ^2-decenoic acid, an antibiotic found in royal jelly. *Science* 130:452-453.

Boch, R. 1956. Die Tänze der Bienen bei nahen und fernen Trachtquellen. *Zeitschrift für Vergleichende Physiologie* 38:136-167.

Boch, R. 1957. Rassenmässige Unterschiede in den Tänzen der Honigbiene (*Apis mellifica* L.). *Zeitschrift für Vergleichende Physiologie* 39:289-320.

Boch, R. and R. A. Morse. 1974. Discrimination of familiar and foreign queens by honeybee swarms. *Annals of the Entomological Society of America* 67:709-711.

Boch, R. and R. A. Morse. 1979. Individual recognition of queens by honeybee swarms. *Annals of the Entomological Society of America.* 72:51-53.

Boch, R. and D. A. Shearer. 1966. Iso-pentyl acetate in stings of honeybees of different ages. *Journal of Apicultural Research* 5:65-70.

Boch, R., D. A. Shearer, and A. Petrasovits. 1970. Efficacies of two alarm substances of the honeybee. *Journal of Insect Physiology* 16:17-24.

Boch, R., D. A. Shearer, and B. C. Stone. 1962. Identification of iso-amyl acetate as an active component in the sting pheromone of the honey bee. *Nature* 195:1018-1020.

Bodenheimer, F. S. 1937. Studies in animal populations. II. Seasonal population trends of the honey-bee. *Quarterly Review of Biology* 12:406-425.

Bonner, J. T. 1965. *Size and Cycle: An Essay on the Structure of Biology.* Princeton University Press, Princeton, N.J.

Böttcher, F. K. 1975. Beiträge zur Kenntnis des Paarungsfluges der Honigbiene. *Apidologie* 6:233-281.

Bozina, K. D. 1961. [How long does a queen live?] *Pchelovodstvo* 38:13. In Russian.

Breed, M. D. 1981. Individual recognition and learning of queen odors by worker honeybees. *Proceedings of the National Academy of Sciences, USA* 78:2635-2637.

Breed, M. D. 1983. Nestmate recognition in honey bees. *Animal Behaviour* 31:86-91.

Brock, T. D. 1967. Life at high temperatures. *Science* 158:1012-1019.

Brokensha, D., H.S.K. Mwaniki, and B. W. Riley. 1972. Beekeeping in Embu district, Kenya. *Bee World* 53:114-123.

Brünnich, K. 1923. A graphic representation of the oviposition of a queen bee. *Bee World* 4:208-210, 223-224.

Bulmer, M. A. 1983. Sex ratio theory in social insects with swarming. *Journal of Theoretical Biology* 100:329-339.

Burleigh, R. and P. Whalley. 1983. On the relative geological age of amber and copal. *Journal of Natural History* 17:919-921.

Butenandt, A. and H. Rembold. 1957. Über den Weiselzellenfuttersaft der Honigbiene. I. Isolierung, Konstitutionsermittlung und Vorkommen der 10-Hydroxy-Δ^2-decensäure. *Hoppe-Seyler's Zeitschrift für Physiologische Chemie* 308:204.

Butler, C. G. 1940. The ages of bees in a swarm. *Bee World* 21:9-10.

Butler, C. G. 1945. The influence of various physical and biological factors of the environment on honeybee activity and nectar concentration and abundance. *Journal of Experimental Biology* 21:5-12.

Butler, C. G. 1954. The method and importance of the recognition by a colony of honeybees (*A. mellifera*) of the presence of its queen. *Transactions of the Royal Entomological Society* (*London*) 105:11-29.

Butler, C. G. 1974. *The World of the Honeybee*. Collins, London.

Butler, C. G. and J. B. Free. 1952. The behaviour of worker honeybees at the hive entrance. *Behaviour* 4:262-292.

Butler, C. G., E. P. Jeffree, and H. Kalmus. 1943. The behaviour of a population of honeybees on an artificial and on a natural crop. *Journal of Experimental Biology* 20:65-73.

Cahill, K. and S. Lustick. 1976. Oxygen consumption and thermoregulation in *Apis mellifera* workers and drones. *Comparative Biochemistry and Physiology* 55A:355-357.

Cale, G. H., Jr. and J. W. Gowen. 1956. Heterosis in the honey bee (*Apis mellifera* L.). *Genetics* 41:292-303.

Callow, R. K. 1963. Chemical and biochemical problems of beeswax. *Bee World* 44:95-101.

Callow, R. K., N. C. Johnston, and J. Simpson. 1959. 10-hydroxy-Δ^2-decenoic acid in the honeybee (*Apis mellifera*). *Experientia* 15:421.

Caron, D. M. 1980. Swarm emergence date and cluster location in honeybees. *American Bee Journal* 119:24-25.

Chadwick, P. C. 1931. Ventilation of the hive. *Gleanings in Bee Culture* 59:356-358.

Chain, B. M. and R. S. Anderson. 1983. Inflammation in insects: the release of a plasmatocyte depletion factor following interaction between bacteria and haemocytes. *Journal of Insect Physiology* 29:1-4.

Chandler, M. T. 1976. The African honeybee—*Apis mellifera adansonii*: the biological basis of its management. In: *Apiculture in Tropical Climates*. E. Crane, ed., pp. 61-68. International Bee Research Association, London.

Charnov, E. L. 1978a. Sex-ratio selection in eusocial Hymenoptera. *American Naturalist* 112:317-326.

Charnov, E. L. 1978b. Evolution of eusocial behavior: offspring choice or parental parasitism? *Journal of Theoretical Biology* 75:451-466.

Charnov, E. L. 1982. *The Theory of Sex Allocation*. Princeton University Press, Princeton, N.J.

Charnov, E. L. and J. R. Krebs. 1973. On clutch size and fitness. *Ibis* 116:217-219.

Charnov, E. L. and W. M. Schaffer. 1973. Life history consequences of natural selection: Cole's result revisited. *American Naturalist* 107:791-793.

Church, N. S. 1960. Heat loss and the temperature of flying insects. II. Heat conductance within the body and its loss by radiation and convection. *Journal of Experimental Biology* 37:186-212.

Čižmárik, J. and I. Matel. 1970. Examination of the chemical composition of propolis. 1. Isolation and identification of the 3,4 dihydroxycinnamic acid (caffeic acid) from propolis. *Experientia* 26:713.

Čižmárik, J. and I. Matel. 1973. Examination of the chemical composition of propolis. 2. Isolation and identification of 4-hydroxy-3-methoxycinnamic acid (ferulic acid) from propolis. *Journal of Apicultural Research* 12:52-54.

Clutton-Brock, T. H. 1977. Some aspects of intraspecific variation in feeding and ranging behaviour in primates. In: *Primate Ecology: Studies of Feeding and Ranging Behaviour in Lemurs, Monkeys and Apes*. T. H. Clutton-Brock, ed., pp. 539-556. Academic Press, London.

Clutton-Brock, T. H. and P. H. Harvey. 1984. Comparative approaches to investigating adaptation. In: *Behavioural Ecology: An Evolutionary Approach*. J. R. Krebs and N. B. Davies, eds., pp. 7-29. Sinauer, Sunderland, Mass.

Cole, B. J. 1981. Dominance hierarchies in *Leptothorax* ants. *Science* 212: 83-84.

Collins, A. M. and M. S. Blum. 1982. Bioassay of compounds derived from the honeybee sting. *Journal of Chemical Ecology* 8:463-470.

Collins, A. M. and M. S. Blum. 1983. Alarm responses caused by newly identified compounds derived from the honeybee sting. *Journal of Chemical Ecology* 9:57-65.

Collins, A. M., T. E. Rinderer, J. R. Harbo, and A. B. Bolten. 1982. Colony defense by Africanized and European honey bees. *Science* 218:72-74.

Combs, G. F. 1972. The engorgement of swarming worker honeybees. *Journal of Apicultural Research* 11:121-128.

Contel, E.P.B., M. A. Mestriner, and E. Martins. 1977. Genetic control and developmental expression of malate dehydrogenase in *Apis mellifera*. *Biochemical Genetics* 15:859-876.

Craig, R. 1979. Parental manipulation, kin selection, and the evolution of altruism. *Evolution* 33:319-334.

Craig, R. 1980. Sex investment ratios in social Hymenoptera. *American Naturalist* 116:311-323.

Craig, R. 1983. Subfertility and the evolution of eusociality by kin selection. *Journal of Theoretical Biology* 100:379-397.

Crane, E. 1975. History of honey. In: *Honey. A Comprehensive Survey*. E. Crane, ed., pp. 439-488. Heinemann, London.

Crane, E. 1983. *The Archaeology of Beekeeping*. Duckworth, London.

Crozier, R. H. 1977. Evolutionary genetics of the Hymenoptera. *Annual Review of Entomology* 22:263-288.

Crozier, R. H. 1979. Genetics of sociality. In: *Social Insects*, Vol. 1. H. R. Hermann, ed., pp. 223-286. Academic Press, New York.

Culliney, T. W. 1983. Origin and evolutionary history of the honeybees *Apis*. *Bee World* 64:29-38.

Dadant, C. P. 1975. Beekeeping equipment. In: *The Hive and the Honeybee*. Dadant and Sons, eds., pp. 303-328. Dadant and Sons, Hamilton, Ill.

Dade, H. A. 1977. *Anatomy and Dissection of the Honeybee*. International Bee Research Association, London.

Daly, H. W. 1975. Identification of Africanized bees by multivariate morphometrics. *Proceedings of the XXVth International Beekeeping Congress, Grenoble*, pp. 356-358.

Darchen, R. 1968. Le travail de la cire et la construction dans la ruche. In: *Traité de biologie de l'abeille*, Vol. 2. R. Chauvin, ed., pp. 241-331. Masson, Paris.

Darwin, C. R. 1859. *On the Origin of Species. A facsimile of the First Edition* [1967]. Harvard University Press, Cambridge, Mass.

Darwin, C. R. 1877. *The Effects of Cross and Self Fertilization in the Vegetable Kingdom*. Appleton, New York.

Dawkins, R. 1976. *The Selfish Gene*. Oxford University Press, Oxford.

Dawkins, R. 1982. *The Extended Phenotype: The Gene as the Unit of Selection*. Freeman, San Francisco.

Dhaliwal, H. S. and P. L. Sharma. 1974. Foraging range of the Indian honeybee. *Journal of Apicultural Research* 13:137-141.

Dold, H., D. H. Du, and S. T. Dziao. 1937. Nachweis antibakterieller hitze- und lichtempfindlicher Hemmungsstoffe (Inhibine) im Naturhonig (Blütenhonig). *Zeitschrift für Hygiene und Infektionskrankheiten* 120: 155-167.

Dunham, W. E. 1929. The influence of external temperature on the hive temperatures during the summer. *Journal of Economic Entomology* 22:798-801.

Dunham, W. E. 1931. A colony of bees exposed to high external temperatures. *Journal of Economic Entomology* 24:606-611.

Dyer, F. C. and J. L. Gould. 1981. Honey bee orientation: a backup system for cloudy days. *Science* 214:1041-1042.

Dyer, F. C. and J. L. Gould. 1983. Honey bee navigation. *American Scientist* 71:587-597.

Dzierzon, J. 1848. *Theorie und Praxis des neuen Bienenfreundes*. (Published by the author.)

Eckert, J. E. 1933. The flight range of the honey bee. *Journal of Agricultural Research* 47:257-285.

Eckert, J. E. 1942. The pollen required by a colony of honeybees. *Journal of Economic Entomology* 35:309-311.

Emerson, A. E. 1960. The evolution of adaptation in population systems. In:

Evolution after Darwin, Vol. 1. Sol Tax, ed., pp. 307-348. University of Chicago Press, Chicago.

Esch, H. 1960. Über die Körpertemperaturen und den Wärmehaushalt von *Apis mellifica*. *Zeitschrift für Vergleichende Physiologie* 43:305-335.

Esch, H. 1964. Über den Zusammenhang zwischen Temperatur, Aktionspotentialen und Thoraxbewegungen bei der Honigbiene (*Apis mellifica* L.). *Zeitschrift für Vergleichende Physiologie* 48:547-551.

Esch, H. 1967. The sound produced by swarming honey bees. *Zeitschrift für Vergleichende Physiologie* 56:408-411.

Esch, H. 1976. Body temperature and flight performance of honey bees in a servo-mechanically controlled wind tunnel. *Journal of Comparative Physiology* 109:265-277.

Esch, H. and J. Bastian. 1968. Mechanical and electrical activity in the indirect flight muscles of the honey bee. *Zeitschrift für Vergleichende Physiologie* 58:429-440.

Esch, H. and J. A. Bastian. 1970. How do newly recruited honeybees approach a food site? *Zeitschrift für Vergleichende Physiologie* 68:175-181.

Evans, H. E. 1977. Extrinsic versus intrinsic factors in the evolution of insect sociality. *Bioscience* 27:613-617.

Farrar, C. L. 1934. Bees must have pollen. *Gleanings in Bee Culture* 62:276-278.

Farrar, C. L. 1936. Influence of pollen reserves on the surviving population of overwintered colonies. *American Bee Journal* 76:452-454.

Fell, R. D. 1977. The study of the biology of queen honey bees (*Apis mellifera* L.) at different times of the year. M.S. thesis, Cornell University.

Fell, R. D., J. T. Ambrose, D. M. Burgett, D. DeJong, R. A. Morse, and T. D. Seeley. 1977. Seasonal cycle of swarming in honey bees (*Apis mellifera* L.). *Journal of Apicultural Research* 16:170-173.

Ferguson, A. W. and J. B. Free. 1980. Queen pheromone transfer within honeybee colonies. *Physiological Entomology* 5:359-366.

Fisher, R. A. 1930, *The Genetical Theory of Natural Selection*. Oxford University Press, Oxford.

Fletcher, D.J.C. 1978. The African bee, *Apis mellifera adansonii*, in Africa. *Annual Review of Entomology* 23:151-171.

Fluri, P., M. Lüscher, H. Wille, and L. Gerig. 1982. Changes in weight of the pharyngeal gland and haemolymph titres of juvenile hormone, protein and vitellogenin in worker honey bees. *Journal of Insect Physiology* 28:61-68.

Fluri, P., H. Wille, L. Gerig, and M. Lüscher. 1977. Juvenile hormone, vitellogenin and haemocyte composition in winter worker honeybees (*Apis mellifera*). *Experientia* 33:1240-1241.

Franks, N. R. and E. Scovell. 1983. Dominance and reproductive success among slave-making worker ants. *Nature* 304:724-725.

Free, J. B. 1960. The behaviour of honeybees visiting the flowers of fruit trees. *Journal of Animal Ecology* 29:385-395.

Free, J. B. 1961. The stimuli releasing the stinging response of honeybees. *Animal Behaviour* 9:193-196.

Free, J. B. 1963. The flower constancy of honeybees. *Journal of Animal Ecology* 32:119-131.

Free, J. B. 1965. The allocation of duties among worker honeybees. *Symposium of the Zoological Society of London* 14:39-59.

Free, J. B. 1967a. The production of drone comb by honeybee colonies. *Journal of Apicultural Research* 6:29-36.

Free, J. B. 1967b. Factors determining the collection of pollen by honeybee foragers. *Animal Behaviour* 15:134-144.

Free, J. B. 1968. The conditions under which foraging honeybees expose their Nasonov gland. *Journal of Apicultural Research* 7:139-145.

Free, J. B. 1970. *Insect Pollination of Crops*. Academic Press, London.

Free, J. B. and J. Simpson. 1963. The respiratory metabolism of honey-bee colonies at low temperatures. *Entomologica Experimentalis et Applicata* 6:234-238.

Free, J. B. and Y. Spencer-Booth. 1958. Observations on the temperature regulation and food consumption of honeybees (*Apis mellifera*). *Journal of Experimental Biology* 35:930-937.

Free, J. B. and Y. Spencer-Booth. 1960. Chill-coma and cold death temperatures of *Apis mellifera*. *Entomologica Experimentalis et Applicata* 3:222-230.

Free, J. B. and Y. Spencer-Booth. 1962. The upper lethal temperatures of honeybees. *Entomologica Experimentalis et Applicata* 5:249-254.

Free, J. B. and I. H. Williams. 1970. Exposure of the Nasonov gland by honeybees (*Apis mellifera*) collecting water. *Behaviour* 37:286-290.

Free, J. B. and I. H. Williams. 1975. Factors determining the rearing and rejection of drones by the honeybee colony. *Animal Behaviour* 23:650-675.

Friedmann, H. 1955. The honey-guides. *Bulletin of the United States National Museum* 208:1-292.

Frisch, K. von. 1923. Über die "Sprache" der Bienen, eine tierpsychologische Untersuchung. *Zoologische Jahrbücher. Abteilung für allgemeine Zoologie und Physiologie der Tiere* 40:1-186.

Frisch, K. von. 1967. *The Dance Language and Orientation of Bees*. Harvard University Press, Cambridge, Mass.

Frisch, K. von. 1974. *Animal Architecture*. Harcourt Brace Jovanovich, New York.

Frisch, K. von and M. Lindauer. 1955. Über die Fluggeschwindigkeit der

Bienen und ihre Richtungsweisung bei Seitenwind. *Naturwissenschaften* 42:377-385.

Frisch, K. von and G. A. Rösch. 1926. Neue Versuche über die Bedeutung von Duftorgan und Pollenduft für die Verständigung im Bienenvolk. *Zeitschrift für Vergleichende Physiologie* 4:1-21.

Fukuda, H., K. Moriga, and K. Sekiguchi. 1969. The weight of crop contents in foraging honeybee workers. *Annotationes Zoologicae Japonenses* 42:80-90.

Galton, D. M. 1971. *Survey of a Thousand Years of Beekeeping in Russia.* Bee Research Association, London.

Gary, N. E. 1960. A trap to quantitatively recover dead and abnormal honeybees from the hive. *Journal of Economic Entomology* 53:782-785.

Gary, N. E. 1962. Chemical mating attractants in the queen honey bee. *Science* 136:773-774.

Gary, N. E. 1963. Observations of mating behaviour in the honeybee. *Journal of Apicultural Research* 2:3-13.

Gary, N. E. 1974. Pheromones that affect the behavior and physiology of honey bees. In: *Pheromones.* M. C. Birch, ed., pp. 200-221. North-Holland, Amsterdam.

Gary, N. E. 1975. Activities and behavior of honey bees. In: *The Hive and the Honeybee.* Dadant and Sons, eds., pp. 185-264. Dadant and Sons, Hamilton, Ill.

Gary, N. E. and J. Marston. 1971. Mating behaviour of drone honeybees with queen models (*Apis mellifera* L.). *Animal Behaviour* 19:299-304.

Gary, N. E. and R. A. Morse. 1962. The events following queen cell construction in honeybee colonies. *Journal of Apicultural Research* 1:3-5.

Gary, N. E., R.F.L. Mau, and W. C. Mitchell. 1972. A preliminary study of honey bee foraging range in macadamia (*Macadamia integrifolia,* Maiden and Betche). *Proceedings of the Hawaiian Entomological Society* 21:205-212.

Gary, N. E., P. C. Witherell, and J. M. Marston. 1972. Foraging range and distribution of honey bees used for carrot and onion pollination. *Environmental Entomology* 1:71-78.

Gary, N. E., P. C. Witherell, and J. M. Marston. 1975. The distribution of foraging honey bees from colonies used for honeydew melon pollination. *Environmental Entomology* 4:277-281.

Gates, B. N. 1914. The temperature of the bee colony. *Bulletin of the United States Department of Agriculture* 96:1-29.

Gauhe, A. 1940. Über ein Glukoseoxydierendes enzym in der pharynxdrüse der Honigbiene. *Zeitschrift für Vergleichende Physiologie* 28:211-253.

Getz, W. M. 1981. Genetically based kin recognition systems. *Journal of Theoretical Biology* 92:209-226.

Getz, W. M., D. Brückner, and T. R. Parisian. 1982. Kin structure and the swarming behavior of the honey bee, *Apis mellifera*. *Behavioral Ecology and Sociobiology* 10:265-270.

Getz, W. M. and K. B. Smith. 1983. Genetic kin recognition: honey bees discriminate between full and half sisters. *Nature* 302:147-148.

Ghisalberti, E. L. 1979. Propolis: a review. *Bee World* 60:59-84.

Gilliam, M. 1971. Microbial sterility of the intestinal contents of the immature honeybee, *Apis mellifera*. *Annals of the Entomological Society of America* 64:315-316.

Gould, J. L. 1975a. Honey bee recruitment: the dance-language controversy. *Science* 189:685-693.

Gould, J. L. 1975b. Communication of distance information by honey bees. *Journal of Comparative Physiology* 104:161-173.

Gould, J. L. 1976. The dance-language controversy. *Quarterly Review of Biology* 51:211-244.

Gould, J. L. 1982. Why do honeybees have dialects? *Behavioral Ecology and Sociobiology* 10:53-56.

Gould, J. L., M. Henerey, and M. C. MacLeod. 1970. Communication of direction by the honey bee. *Science* 169:544-554.

Groot, A. P. de and S. Voogd. 1954. On the ovary development in queenless worker bees (*Apis mellifica* L.). *Experientia* 10:384-385.

Guy, R. D. 1972a. Commercial beekeeping with African bees. *Bee World* 53:14-22.

Guy, R. D. 1972b. The honey hunters of Southern Africa. *Bee World* 53:159-166.

Guy, R. D. 1976. Commercial beekeeping with *Apis mellifera adansonii* in intermediate and movable-frame hives. In: *Apiculture in Tropical Climates*. E. Crane, ed., pp. 31-37. International Bee Research Association, London.

Habermann, E. 1971. Chemistry, pharmacology, and toxicology of bee, wasp, and hornet venoms. In: *Venomous Animals and Their Venoms*, Vol. 3. *Venomous Invertebrates*. W. Bücherl and E. E. Buckley, eds., pp. 61-94. Academic Press, New York.

Habermann, E. 1972. Bee and wasp venoms. *Science* 177:314-322.

Hamilton, W. D. 1964. The genetical evolution of social behavior. *Journal of Theoretical Biology* 7:1-52.

Hamilton, W. D. 1967. Extraordinary sex ratios. *Science* 156:477-488.

Hamilton, W. D. 1972. Altruism and related phenomena, mainly in social insects. *Annual Review of Ecology and Systematics* 3:193-232.

Hamilton, W. D. 1975. Gamblers since life began: barnacles, aphids, elms. *Quarterly Review of Biology* 50:175-180.

Hamilton, W. D. 1979. Wingless and fighting males in fig wasps and other insects. In: *Sexual Selection and Reproductive Competition in Insects*.

M. S. Blum and N. A. Blum, eds., pp. 167-220. Academic Press, New York.

Hannson, Å. 1975. Lauterzeugung und Lautauffassungvermögen der Biene. *Opuscula Entomologica. Supplement* 6:1-124.

Harborne, J. B. 1982. *Introduction to Ecological Biochemistry.* Academic Press, London.

Haydak, M. H. 1935. Brood rearing by honeybees confined to a pure carbohydrate diet. *Journal of Economic Entomology* 28:657-660.

Hazelhoff, E. H. 1954. Ventilation in a bee-hive during summer. *Physiologia Comparata et Oecologia* 3:343-364.

Heinrich, B. 1977. Why have some animals evolved to regulate a high body temperature? *American Naturalist* 111:623-640.

Heinrich, B. 1978. The economics of insect sociality. In: *Behavioural Ecology: An Evolutionary Approach.* J. R. Krebs and N. B. Davies, eds., pp. 97-128. Sinauer, Sunderland, Mass.

Heinrich, B. 1979a. *Bumblebee Economics.* Harvard University Press, Cambridge, Mass.

Heinrich, B. 1979b. Thermoregulation of African and European honeybees during foraging, attack, and hive exits and returns. *Journal of Experimental Biology* 80:217-229.

Heinrich, B. 1980. Mechanisms of body-temperature regulation in honeybees, *Apis mellifera. Journal of Experimental Biology* 85:61-87.

Heinrich, B. 1981a. The mechanisms and energetics of honeybee swarm temperature regulation. *Journal of Experimental Biology* 91:25-55.

Heinrich, B. 1981b. Ecological and evolutionary perspectives. In: *Insect Thermoregulation.* B. Heinrich, ed., pp. 236-302. Wiley, New York.

Heinrich, B. 1981c. Energetics of honeybee swarm thermoregulation. *Science* 212:565-566.

Heinrich, B. 1983. Insect foraging energetics. In: *Handbook of Experimental Pollination Biology.* C. E. Jones and R. J. Little, eds., pp. 187-214. Van Nostrand Reinhold, New York.

Heran, H. 1952. Untersuchungen über den Temperatursinn der Honigbiene (*Apis mellifica*) unter besonderer Berücksichtigung der Wahrnehmung strahlender Wärme. *Zeitschrift für Vergleichende Physiologie* 34:179-206.

Herreid, C. F. and B. Kessel. 1967. Thermal conductance in birds and mammals. *Comparative Biochemistry and Physiology* 21:405-414.

Hess, W. R. 1926. Die Temperaturregulierung im Bienenvolk. *Zeitschrift für Vergleichende Physiologie* 4:465-487.

Heussner, A. and M. Roth. 1963. Consommation d'oxygène de l'abeille à différentes températures. *Compte Rendu de l'Académie des Sciences, Paris* 256:284-285.

Heussner, A. and T. Stussi. 1964. Métabolisme énergétique de l'abeille isolée. Son rôle dans la thermorégulation de la ruche. *Insectes Sociaux* 11: 239-266.

Himmer, A. 1927. Ein Beitrag zur Kenntnis des Wärmehaushalts im Nestbau sozialer Hautflüger. *Zeitschrift für Vergleichende Physiologie* 5:375-389.

Hirschfelder, H. 1951. Quantitative Untersuchungen zum Polleneintragen der Bienenvölker. *Zeitschrift für Bienenforschung* 1:67-77.

Hoefer, I. and M. Lindauer. 1975. Das Lernverhalten zweier Bienenrassen unter veränderten Orientierungbedingungen. *Journal of Comparative Physiology* 99:119-138.

Hölldobler, B. and C. D. Michener. 1980. Mechanisms of identification and discrimination in social Hymenoptera. In: *Evolution of Social Behavior: Hypotheses and Empirical Tests.* H. Markl, ed., pp. 35-58. Verlag Chemie, Weinheim.

Holmes, W. G. and P. W. Sherman. 1982. The ontogeny of kin recognition in two species of ground squirrels. *American Zoologist* 22:491-517.

Holmes, W. G. and P. W. Sherman. 1983. Kin recognition in animals. *American Scientist* 71:46-55.

Horn, H. S. 1978. Optimal tactics of reproduction and life-history. In: *Behavioural Ecology: An Evolutionary Approach.* J. R. Krebs and N. B. Davies, eds., pp. 411-429. Sinauer, Sunderland, Mass.

Horstmann, H.-J. 1965. Einige biochemischen Überlegungen zur Bildung von Bienenwachs aus Zucker. *Zeitschrift für Bienenforschung* 8:125-128.

Huber, F. 1792. *Nouvelles observations sur les abeilles*, adressées à M. Charles Bonnet. Barde, Manget and Co., Geneva. [1926. English translation by C. P. Dadant, as *New Observations upon Bees*. American Bee Journal, Hamilton, Ill.]

Huber, F. 1814. *Nouvelles observations sur les abeilles*. II. Paschoud, Geneva. [1926. English translation by C. P. Dadant, as *New Observations upon Bees*. American Bee Journal, Hamilton, Ill.]

Janscha, A. 1771. *Abhandlung von Schwärmen der Bienen.* Vienna, Kurzbock.

Janzen, D. H. 1979. New horizons in the biology of plant defenses. In: *Herbivores: Their Interaction with Secondary Plant Metabolites.* G. A. Rosenthal and D. H. Janzen, eds., pp. 331-350. Academic Press, New York.

Janzen, D. H. 1981. Evolutionary physiology of personal defense. In: *Physiological Ecology: An Evolutionary Approach to Resource Use.* C. R. Townsend and P. Calow, eds., pp. 145-164. Sinauer, Sunderland, Mass.

Jarman, P. J. 1982. Prospects for interspecific comparison in sociobiology.

In: *Current Problems in Sociobiology*. King's College Sociobiology Group, eds., pp. 323-342. Cambridge University Press, Cambridge.

Jay, S. C. 1959. Factors affecting the laboratory rearing of honeybee larvae (*Apis mellifera* L.). M.S. thesis, University of Toronto.

Jay, S. C. 1968. Factors influencing ovary development of worker honeybees under natural conditions. *Canadian Journal of Zoology* 46:345-347.

Jay, S. C. 1970. The effect of various combinations of immature queen and worker bees on the ovary development of worker honeybees in colonies with and without queens. *Canadian Journal of Zoology* 48:169-173.

Jay, S. C. and D. H. Jay. 1976. The effect of various types of brood comb on the ovary development of worker honeybees. *Canadian Journal of Zoology* 54:1724-1726.

Jaycox, E. R. and S. G. Parise. 1980. Homesite selection by Italian honey bee swarms, *Apis mellifera ligustica* (Hymenoptera: Apidae). *Journal of the Kansas Entomological Society* 53:171-178.

Jaycox, E. R. and S. G. Parise. 1981. Homesite selection by swarms of black-bodied honeybees. *Apis mellifera caucasica* and *A. m. carnica* (Hymenoptera: Apidae). *Journal of the Kansas Entomological Society* 54:697-703.

Jeanne, R. L. 1979. A latitudinal gradient of rates of ant predation. *Ecology* 60:1211-1224.

Jean-Prost, P. 1956. Quelques points de la biologie des abeilles en Provence. *Revue Française d'Apiculture* 3:1558-1561.

Jean-Prost, P. 1958. Resumé des observations sur le vol nuptial des reines d'abeilles. *Proceedings of the XVIIth International Beekeeping Congress, Rome*, pp. 404-408.

Jeffree, E. P. 1951. The swarming period in Wiltshire. *Wiltshire Beekeepers Gazette* 74:2-3.

Jeffree, E. P. 1955. Observations on the decline and growth of honey bee colonies. *Journal of Economic Entomology* 48:723-726.

Jeffree, E. P. 1956. Winter brood and pollen in honey bee colonies. *Insectes Sociaux* 3:417-422.

Johnson, D. L. and A. M. Wenner. 1970. Recruitment efficiency in honey bees: studies on the role of olfaction. *Journal of Apicultural Research* 9:13-18.

Jongbloed, J. and C.A.G. Wiersma. 1934. Der Stoffwechsel der Honigbiene während des Fliegens. *Zeitschrift für Vergleichende Physiologie* 21:519-533.

Josephson, R. K. 1981. Temperature and the mechanical performance of insect muscle. In: *Insect Thermoregulation*. B. Heinrich, ed., pp. 20-44. Wiley, New York.

Kalmus, H. and C. R. Ribbands. 1952. The origin of the odours by which honeybees distinguish their companions. *Proceedings of the Royal Society* (B)140:50-59.

Kamil, A. C. and T. D. Sargeant. 1981. *Foraging Behavior: Ecological, Ethological, and Psychological Approaches.* Garland STPM, New York.

Kammer, A. E. 1981. Physiological mechanisms of thermoregulation. In: *Insect Thermoregulation.* B. Heinrich, ed., pp. 115-158. Wiley, New York.

Kammer, A. E. and B. Heinrich. 1974. Metabolic rates related to muscle activity in bumblebees. *Journal of Experimental Biology* 61:219-227.

Kammer, A. E. and B. Heinrich. 1978. Insect flight metabolism. *Advances in Insect Physiology* 13:133-228.

Kefuss, J. A. 1978. Influence of photoperiod on the behaviour and brood rearing activities of the honeybee. *Journal of Apicultural Research* 17:137-151.

Kerr, W. E. 1967. Multiple alleles and genetic load in bees. *Journal of Apicultural Research* 6:61-64.

Kerr, W. E. 1969. Some aspects of the evolution of social bees. *Evolutionary Biology* 3:119-175.

Kerr, W. E. and N. J. Hebling. 1964. Influence of the weight of worker bees on division of labor. *Evolution* 18:267-270.

Kerr, W. E., M. R. Martinho, and L. S. Gonçalves. 1980. Kinship selection in bees. *Revista Brasileira de Genética* 3:339-344.

Kerr, W. E., R. Zucchi, J. T. Nakadaira, and J. E. Butolo. 1962. Reproduction in the social bees. *Journal of the New York Entomological Society* 70: 265-270.

Kiechle, H. 1961. Die soziale Regulation der Wassersammeltätigkeit im Bienenstaat und deren physiologische Grundlage. *Zeitschrift für Vergleichende Physiologie* 45:154-192.

Kigatiira, K. I. 1984. Bees and beekeeping in Kenya. *Bee World* 65:74-80.

Kluger, M. J. 1979. *Fever, Its Biology, Evolution, and Function.* Princeton University Press, Princeton, N.J.

Knaffl, H. 1953. Über die Flugweite und Entfernungsmeldung der Bienen. *Zeitschrift für Bienenforschung* 2:131-140.

Koch, H. G. 1967. Der Jahresgang der Nektartracht von Bienenvölkern als Ausdruck der Witterungssingularitäten und Trachtverhältnisse. *Zeitschrift für Angewandte Meteorologie* 5:206-216.

Koeniger, N. and G. Koeniger. 1980. Observations and experiments on migration and dance communication of *Apis dorsata* in Sri Lanka. *Journal of Apicultural Research* 19:21-34.

Koeniger, G., N. Koeniger, and M. Fabritius. 1979. Some detailed observations of mating in the honeybee. *Bee World* 60:53-57.

Koeniger, N. and H. J. Veith. 1983. Glyceryl-1,2-dioleate-3-palmitate, a brood pheromone of the honey bee (*Apis mellifera* L.). *Experientia* 39:1051-1052.

Koeniger, N. and G. Vorwohl. 1979. Competition for food among four sympatric species of Apini in Sri Lanka (*Apis dorsata, Apis cerana, Apis florea* and *Trigona iridipennis*). *Journal of Apicultural Research* 18: 95-109.

Köhler, F. 1955. Wache und Volksduft im Bienenstaat. *Zeitschrift für Bienenforschung* 3:57-63.

Koltermann, R. 1969. Lern- und Vergessensprozesse bei der Honigbiene — aufgezeigt anhand von Duftdressuren. *Zeitschrift für Vergleichende Physiologie* 63:310-334.

Koltermann, R. 1973. Rassen bzw. artspezifische Duftbewertung bei der Honigbiene und ökologische Adaptation. *Journal of Comparative Physiology* 85:327-360.

Korst, P.J.A.M. and H.H.W. Velthuis. 1982. The nature of trophallaxis in honeybees. *Insectes Sociaux* 29:209-221.

Kosmin, N. P., W. W. Alpatov, and M. S. Resnitschenko. 1932. Zur Kenntnis des Gaswechsels und Energieverbrauchs der Biene in Beziehung zu deren Aktivität. *Zeitschrift für Vergleichende Physiologie* 17:408-422.

Krebs, J. R. 1978. Optimal foraging: decision rules for predators. In: *Behavioural Ecology: An Evolutionary Approach*. J. R. Krebs and N. B. Davies, eds., pp. 23-63. Sinauer, Sunderland, Mass.

Krebs, J. R. 1980. Foraging strategies and their social significance. In: *Handbook of Behavioural Neurobiology*, Vol. 3. P. Marler and J. Vandenbergh, eds., pp. 225-270. Plenum Press, New York.

Kronenberg, F. and H. C. Heller. 1982. Colonial thermoregulation in honey bees (*Apis mellifera*). *Journal of Comparative Physiology* 148:65-76.

Kropáčová, S. and H. Haslbachová. 1969. The development of ovaries in worker honeybees in a queenright colony. *Journal of Apicultural Research* 8:57-64.

Kropáčová, S. and H. Haslbachová. 1970. The development of ovaries in worker honeybees in queenright colonies examined before and after swarming. *Journal of Apicultural Research* 9:65-70.

Kropáčová, S. and H. Haslbachová. 1971. The influence of queenlessness and of unsealed brood on the development of ovaries in worker honeybees. *Journal of Apicultural Research* 10:57-61.

Kulinčević, J. M. and W. C. Rothenbuhler. 1975. Selection for resistance and susceptibility to hairless-black syndrome in the honey bee. *Journal of Invertebrate Pathology* 25:289-295.

Lacher, V. 1964. Elektrophysiologische Untersuchung an einzelnen Rezeptoren für Geruch, Luftfeuchtigkeit und Temperatur auf den Antennen der

Arbeitsbienen und der Drohne. *Zeitschrift für Vergleichende Physiologie* 48:587-623.

Lack, D. 1954. *The Natural Regulation of Animal Numbers.* Oxford University Press, Oxford.

Lack, D. 1966. *Population Studies of Birds.* Oxford University Press, Oxford.

Lack, D. 1968. *Ecological Adaptations for Breeding in Birds.* Methuen, London.

Laidlaw, H. H., Jr. 1944. Artificial insemination of the queen bee (*Apis mellifera* L.): Morphological basis and results. *Journal of Morphology* 74:429-465.

Laidlaw, H. H., F. P. Gomes, and W. E. Kerr. 1956. Estimations of the number of lethal alleles in a panmictic population of *Apis mellifera* L. *Genetics* 41:179-188.

Lavie, P. 1968. Les substances antibiotiques dans la colonie d'abeilles. In: *Traité de biologie de l'abeille*, Vol. 3. R. Chauvin, ed., pp. 2-115. Masson, Paris.

Lensky, Y. and Y. Slabezki. 1981. The inhibitory effect of the queen bee (*Apis mellifera* L.) foot-print pheromone on the construction of swarming queen cups. *Journal of Insect Physiology* 27:313-323.

Levchenko, I. A. 1959. [The distance bees fly for nectar]. *Pchelovodstvo* 36:37-38. In Russian.

Levin, M. D. and S. Glowska-Konopacka. 1963. Responses of foraging honeybees in alfalfa to increasing competition from other colonies. *Journal of Apicultural Research* 2:33-42.

Lewontin, R. C. 1970. The units of selection. *Annual Review of Ecology and Systematics* 1:1-18.

Lindauer, M. 1948. Über die Einwirkung von Duft- und Geschmacksstoffen sowie anderer Faktoren auf die Tänze der Bienen. *Zeitschrift für Vergleichende Physiologie* 31:348-412.

Lindauer, M. 1952. Ein Beitrag zur Frage der Arbeitsteilung im Bienenstaat. *Zeitschrift für Vergleichende Physiologie* 34:299-345.

Lindauer, M. 1954. Temperaturregulierung und Wasserhaushalt im Bienenstaat. *Zeitschrift für Vergleichende Physiologie* 36:391-432.

Lindauer, M. 1955. Schwarmbienen auf Wohnungssuche. *Zeitschrift für Vergleichende Physiologie* 37:263-324.

Lindauer, M. 1956. Über die Verständigung bei indischen Bienen. *Zeitschrift für Vergleichende Physiologie* 38:521-557.

Lindauer, M. 1959. Angeborene und erlernte Komponenten in der Sonnenorientierung der Bienen. *Zeitschrift für Vergleichende Physiologie* 42:43-62.

Lindauer, M. 1967. Recent advances in bee communication and orientation. *Annual Review of Entomology* 12:439-470.

Lindauer, M. 1970. Lernen und Gedächtnis — Versuche an der Honigbiene. *Naturwissenschaften* 57:463-467.

Lindenfelser, L. A. 1967. Antimicrobial activity of propolis. *American Bee Journal* 107:90-92, 130-131.

Little, H. F. 1962. Reactions of the honey bee, *Apis mellifera* L., to artificial sounds and vibrations of known frequencies. *Annals of the Entomological Society of America* 55:82-89.

Lotmar, R. 1951. Gewichtsbestimmungen bei gesunden und nosemakranken Bienen. *Zeitschrift für Vergleichende Physiologie* 33:195-206.

Louveaux, J. 1958. Recherches sur la récolte du pollen par les abeilles (*Apis mellifica* L.). *Annales de l'Abeille* 1:113-188, 197-221.

Louveaux, J. 1973. The acclimatization of bees to a heather region. *Bee World* 54:105-111.

Lyr, H. 1961. Hemmungsanalytische Untersuchungen an einigen Ektoenzymen holzzerstörender Pilze. *Enzymologia* 23:231-248.

Maa, T. 1953. An inquiry into the systematics of the tribus Apidini or honeybees (Hym.). *Treubia* 21:525-640.

Macevicz, S. 1979. Some consequences of Fisher's sex ratio principle for social Hymenoptera that reproduce by colony fission. *American Naturalist* 113:363-371.

Mackensen, O. and W. C. Roberts. 1948. *A Manual for the Artificial Insemination of Queen Bees*. United States Department of Agriculture, Bureau of Entomology and Plant Quarantine, Bulletin ET-250.

Martin, H. and M. Lindauer. 1966. Sinnesphysiologische Leistungen beim Wabenbau der Honigbiene. *Zeitschrift für Vergleichende Physiologie* 53:372-404.

Martin, P. 1963. Die Steuerung der Volksteilung beim Schwärmen der Bienen. Zugleich ein Beitrag zum Problem der Wanderschwärme. *Insectes Sociaux* 10:13-42.

Martins, E., M. A. Mestriner, and E.P.B. Contel. 1977. Alcohol dehydrogenase polymorphism in *Apis mellifera*. *Biochemical Genetics* 15:357-366.

Maschwitz, U. W. 1964. Gefahrenalarmstoffe und Gefahrenalarmierung bei sozialen Hymenopteren. *Zeitschrift für Vergleichende Physiologie* 47:596-655.

Maschwitz, U.W.J. and W. Kloft. 1971. Morphology and function of the venom apparatus of insects—bees, wasps, ants, and caterpillars. In: *Venomous Animals and Their Venoms*, Vol. 3. *Venomous Invertebrates*. W. Bücherl and E. E. Buckley, eds., pp. 1-60. Academic Press, New York.

Maul, V. 1969. The cause of the hybridization barrier between *Apis mellifera* L. and *Apis cerana* Fabr. (= syn. *A. indica* Fabr.). 2. Egg fertilization and embryonic development. *Proceedings of the XXIInd International Beekeeping Congress, Munich*, p. 561.

Maurizio, A. 1934. Über die Kalkbrut (Pericystis–Mykose) der Bienen. *Archiv für Bienenkunde* 15:165-193.

Maurizio, A. 1953. Weitere Untersuchungen an Pollenhöschen. Beitrag zur Erfassung der Pollentractverhältnisse in verschiedenen Gegenden der Schweiz. *Schweizerische Bienen-zeitung. Beihefte* 2:485-556.

Mautz, D. 1971. Der Kommunikationseffekt der Schwänzeltänze bei *Apis mellifica carnica* (Pollm.). *Zeitschrift für Vergleichende Physiologie* 72:197-220.

May, M. L. 1976. Warming rates as a function of body size in periodic endotherms. *Journal of Comparative Physiology* 111:55-70.

McCleskey, C. S. and R. M. Melampy. 1934. Bactericidal properties of royal jelly of the honeybee. *Journal of Economic Entomology* 32:581-587.

McLellan, A. R. 1977. Honeybee colony weight as an index of honey production and nectar flow: a critical evaluation. *Journal of Applied Ecology* 14:401-408.

Menzel, R., J. Erber, and T. Masuhr. 1974. Learning and memory in the honey bee. In: *Experimental Analysis of Insect Behaviour*. L. Barton Browne, ed., pp. 195-217. Springer-Verlag, New York.

Menzel, R., H. Freudel, and U. Rühl. 1973. Rassenspezifische Unterschiede im Lernverhalten der Honigbiene (*Apis mellifica* L.). *Apidologie* 4:1-24.

Merrill, J. H. 1924. Observations on brood rearing. *American Bee Journal* 64:337-338.

Mestriner, M. A. and E.P.B. Contel. 1972. The P-3 and EST loci in the honey bee, *Apis mellifera*. *Genetics* 72:733-738.

Meyer, W. 1956a. Arbeitsteilung im Bienenschwarm. *Insectes Sociaux* 3: 303-324.

Meyer, W. 1956b. "Propolis bees" and their activities. *Bee World* 37: 25-36.

Michener, C. D. 1964. Evolution of the nests of bees. *American Zoologist* 4:227-239.

Michener, C. D. 1973. The Brazilian honeybee. *Bioscience* 23:523-527.

Michener, C. D. 1974. *The Social Behavior of the Bees: A Comparative Study*. Harvard University Press, Cambridge, Mass.

Michener, C. D. 1975. The Brazilian bee problem. *Annual Review of Entomology* 20:399-416.

Michener, C. D. 1979. Biogeography of the bees. *Annals of the Missouri Botanical Garden* 66:277-347.

Michener, C. D. 1982. The Africanized honey bee. Introduction. In: *Social Insects in the Tropics*, Vol. 1. P. Jaisson, ed., pp. 205-207. Université Paris-Nord, Paris.

Milum, V. G. 1930. Variations in time of development of the honeybee. *Journal of Economic Entomology* 23:441-447.

Milum, V. G. 1955. Honey bee communication. *American Bee Journal* 95:97-104.

Milum, V. G. 1956. An analysis of twenty years of honeybee colony weight changes. *Journal of Economic Entomology* 49:735-738.

Mitchell, C. 1970. Weights of workers and drones. *American Bee Journal* 110:468-469.

Mitchener, A. V. 1948. The swarming season for honeybees in Manitoba. *Journal of Economic Entomology* 41:646.

Mitchener, A. V. 1955. Manitoba nectar flows 1924-1954, with particular reference to 1947-1954. *Journal of Economic Entomology* 48:514-518.

Morse, R. A. 1978. *Honey Bee Pests, Predators, and Diseases.* Cornell University Press, Ithaca, N.Y.

Murray, L. and E. P. Jeffree. 1955. Swarming in Scotland. *Scottish Beekeeper* 31:96-98.

Murrell, D. C. and W. T. Nash. 1981. Nectar secretion by toria (*Brassica campestris* L. v. *toria*) and foraging behaviour of three *Apis* species in Bangladesh. *Journal of Apicultural Research* 20:34-38.

Naile, F. 1976. *America's Master of Bee Culture: The Life of L. L. Langstroth.* Cornell University Press, Ithaca, N.Y.

Nedel, J. O. 1960. Morphologie und Physiologie der Mandibeldrüse einiger Bienen-Arten (Apidae). *Zeitschrift für Morphologie und Ökologie der Tiere* 49:139-183.

Neuhaus, W. and R. Wohlgemuth. 1960. Über das Fächeln der Bienen und dessen Verhältnis zum Fliegen. *Zeitschrift für Vergleichende Physiologie* 43:615-641.

Neville, A. C. 1965. Energy economy in insect flight. *Science Progress* 53:203-219.

Nightingale, J. M. 1976. Traditional beekeeping among Kenya tribes, and methods proposed for improvement and modernisation. In: *Apiculture in Tropical Climates.* E. Crane, ed., pp. 15-22. International Bee Research Association, London.

Nitschmann, J. 1957. Die Füllung der Rektalblase von *Apis mellifera* L. im Winter (Hym. Apidae). *Deutsche Entomologische Zeitschrift* 4:143-171.

Nolan, W. J. 1925. The brood-rearing cycle of the honeybee. *Bulletin of the United States Department of Agriculture* 1349:1-56.

Norton-Griffiths, M., D. Herlocker, and L. Pennycuick. 1975. The patterns of rainfall in the Serengeti ecosystem, Tanzania. *East Africa Wildlife Journal* 13:347-374.

Nowogrodzki, R. 1981. Regulation of the number of foragers on a constant food source by honey bee colonies. M.S. thesis, Cornell University.

Nowogrodzki, R. 1983. Individual differences and division of labor in honey bees. Ph.D. thesis, Cornell University.

Nunamaker, R. A. and W. T. Wilson. 1981. Comparison of MDH allozyme patterns in the African honey bee (*Apis mellifera adansonii* L.) and the Africanized populations of Brazil. *Journal of the Kansas Entomological Society* 54:704-710.

Núñez, J. A. 1966. Quantitative Beziehungen zwischen den Eigenschaften von Futterquellen und den Verhalten von Sammelbienen. *Zeitschrift für Vergleichende Physiologie* 53:142-164.

Núñez, J. A. 1970. The relationship between sugar flow and foraging and recruiting behaviour of honey bees (*Apis mellifera* L.). *Animal Behaviour* 18:527-538.

Núñez, J. A. 1979a. Time spent on various components of foraging activity: comparison between European and Africanized honeybees in Brazil. *Journal of Apicultural Research* 18:110-115.

Núñez, J. A. 1979b. Comparative study of thermoregulation between European and Africanized *Apis mellifera* in Brazil. *Journal of Apicultural Research* 18:116-121.

Núñez, J. A. 1982. Honeybee foraging strategies at a food source in relation to its distance from the hive and the rate of sugar flow. *Journal of Apicultural Research* 21:139-150.

Nye, W. P. and O. Mackensen. 1968. Selective breeding of honeybees for alfalfa pollination: Fifth generation and backcrosses. *Journal of Apicultural Research* 7:21-27.

O'Connor, R. and M. L. Peck. 1978. Venoms of Apidae. In: *Handbook of Experimental Pharmacology*, Vol. 48. S. Bettini, ed., pp. 613-659. Springer-Verlag, Heidelberg.

Oertel, E. 1944. Variation in the sugar concentration of some southern nectars. *Journal of Economic Entomology* 37:525-527.

Oertel, E. 1958. Colony gains and losses at two locations in Louisiana. *American Bee Journal* 98:62-63.

Oettingen-Spielberg, T. zu. 1949. Über das Wesen der Suchbienen. *Zeitschrift für Vergleichende Physiologie* 31:454-489.

Olifir, V. N. 1969. [Territories of food foraging and nectar collection.] *Pchelovodstvo* 89:18-19. In Russian.

Opfinger, E. 1949. Zur Psychologie der Duftdressuren bei Bienen. *Zeitschrift für Vergleichende Physiologie* 31:441-453.

Orians, G. H. 1980. *Some Adaptations of Marsh-Nesting Blackbirds*. Princeton University Press, Princeton, N.J.

Orians, G. H. and N. E. Pearson. 1979. On the theory of central place foraging. In: *Analysis of Ecological Systems*. D. J. Horn, R. D. Mitchell, and G. R. Stairs, eds., pp. 155-177. Ohio State University Press, Columbus, Ohio.

Oster, G. F. and E. O. Wilson. 1978. *Caste and Ecology in the Social Insects.* Princeton University Press, Princeton, N.J.

Otis, G. W. 1982a. Weights of worker honeybees in swarms. *Journal of Apicultural Research* 21:88-92.

Otis, G. W. 1982b. Population biology of the Africanized honey bee. In: *Social Insects in the Tropics*, Vol. 1. P. Jaisson, ed., pp. 209-219. Université Paris-Nord, Paris.

Otis, G. W., M. L. Winston, and O. R. Taylor, Jr. 1981. Engorgement and dispersal of Africanized honeybee swarms. *Journal of Apicultural Research* 20:3-12.

Owens, C. D. 1971. The thermology of wintering honey bee colonies. *Technical Bulletin, United States Department of Agriculture* 1429:1-32.

Page, R. E., Jr. 1980. The evolution of multiple mating behavior by honey bee queens (*Apis mellifera* L.). *Genetics* 96:263-273.

Page, R. E., Jr. 1981. Protandrous reproduction in honey bees (*Apis mellifera* L.). *Environmental Entomology* 10:359-362.

Page, R. E., Jr. 1982. The seasonal occurrence of honey bee swarms in north-central California. *American Bee Journal* 121:266-272.

Page, R. E., Jr., R. B. Kimsey, and H. H. Laidlaw, Jr. 1984. Migration and dispersal of spermatozoa in spermathecae of queen honeybees (*Apis mellifera* L.). *Experientia* 40:182-184.

Page, R. E., Jr. and R. A. Metcalf. 1982. Multiple mating, sperm utilization, and social evolution. *American Naturalist* 119:263-281.

Pager, H. 1971. *Ndedema.* Akademische Druck- und Verlagsanstalt, Graz, Austria.

Pager, H. 1973. Rock paintings in Southern Africa showing bees and honey hunting. *Bee World* 54:61-68.

Pain, J. 1954. Sur l'ectohormone des reines d'abeilles. *Compte Rendu de l'Académie des Sciences, Paris* 239:1869-1870.

Park, O. W. 1928. Time factors in relation to the acquisition of food by the honeybee. *Research Bulletin of the Iowa Agricultural Experiment Station* 108:183-225.

Park, O. W. 1937. Testing for resistance to American foulbrood in honeybees. *Journal of Economic Entomology* 30:504-512.

Park, O. W. 1949. Activities of honey bees. In: *The Hive and the Honeybee.* R. A. Grout, ed., pp. 79-152. Dadant and Sons, Hamilton, Ill.

Park, O. W. 1953. Results of Iowa's honeybee disease resistance program, 1935-1949. *Proceedings of the Iowa Academy of Sciences* 60:707-715.

Park, O. W., F. C. Pellett, and F. B. Paddock. 1937. Disease resistance and American foulbrood: Results of second season of cooperative experiment. *American Bee Journal* 77:20-25, 34.

Parker, R. L. 1926. The collection and utilization of pollen by the honeybee. *Cornell University Agricultural Experiment Station Memoir* 98:1-55.

Peer, D. F. 1957. Further studies on the mating range of the honey bee, *Apis mellifera* L. *The Canadian Entomologist* 89:108-110.

Peer, D. F. and C. L. Farrar. 1956. The mating range of the honey bee. *Journal of Economic Entomology* 49:254-256.

Pellett, F. C. 1938. *History of American Beekeeping*. Collegiate Press, Ames, Iowa.

Pennycuick, C. J. 1979. Energy costs of locomotion and the concept of foraging radius. In: *Serengeti: Dynamics of an Ecosystem*. A.R.E. Sinclair and M. Norton-Griffiths, eds., pp. 164-184. University of Chicago Press, Chicago.

Percival, M. S. 1965. *Floral Biology*. Pergamon, Oxford.

Peters, R. H. 1983. *The Ecological Implications of Body Size*. Cambridge University Press, Cambridge.

Phillips, E. F. 1915. *Beekeeping*. Macmillan, New York.

Phillips, E. F. and G. S. Demuth. 1914. The temperature of the honeybee cluster in winter. *Bulletin of the United States Department of Agriculture* 93:1-16.

Pleasants, J. M. and M. L. Zimmerman. 1979. Patchiness in the dispersion of nectar resources: evidence for hot and cold spots. *Oecologia* 41: 283-288.

Portugal Araújo, V. de. 1971. The Central African bee in South America. *Bee World* 52:116-121.

Prica, M. 1938. Über die bactericide Wirkung des Naturhonigs. *Zeitschrift für Hygiene und Infektionskrankheiten* 120:437-443.

Pyke, G. H. 1981. Optimal foraging in nectar-feeding animals and coevolution with their plants. In: *Foraging Behavior: Ecological, Ethological, and Psychological Approaches*. A. C. Kamil and T. D. Sargent, eds., pp. 19-38, Garland STPM, New York.

Pyke, G. H., H. R. Pulliam, and E. L. Charnov. 1977. Optimal foraging: a selective review of theory and tests. *Quarterly Review of Biology* 52: 137-154.

Queeny, E. M. 1952. The Wandorobo and the honey guide. *Natural History* 61:392-396.

Ratner, S. and S. B. Vinson. 1983. Phagocytosis and encapsulation: cellular immune responses in the Arthropoda. *American Zoologist* 23:185-194.

Ribbands, C. R. 1949. The foraging method of individual honeybees. *Journal of Animal Ecology* 18:47-66.

Ribbands, C. R. 1951. The flight range of the honeybee. *Journal of Animal Ecology* 20:220-226.

Ribbands, C. R. 1952. Division of labour in the honeybee community. *Proceedings of the Royal Society* (B)140:32-43.

Ribbands, C. R. 1953. *The Behaviour and Social Life of Honeybees*. Bee Research Association, London.

Ribbands, C. R. 1954. The defence of the honeybee community. *Proceedings of the Royal Society* (B)142:514-524.

Richards, K. W. 1973. Biology of *Bombus polaris* Curtis and *B. hyperhoreus* Schönherr at Lake Hazen, Northwest Territories (Hymenoptera: Bombini). *Quaestiones Entomologicae* 9:115-157.

Riches, H.R.C. 1982. Hypersensitivity to bee venom. *Bee World* 63:7-22.

Riemann, G. 1958. Zahlen, die zu denken geben. *Die Bienenzucht* 11:121.

Rinderer, T. E. 1982. Regulated nectar harvesting by the honeybee. *Journal of Apicultural Research* 21:74-87.

Rinderer, T. E. and J. R. Baxter. 1978. Effect of empty comb on hoarding behavior and honey production of the honey bee. *Journal of Economic Entomology* 71:757-759.

Rinderer, T. E., A. B. Bolten, J. R. Harbo, and A. M. Collins. 1982. Hoarding behavior of European and Africanized honey bees (Hymenoptera: Apidae). *Journal of Economic Entomology* 75:714-715.

Rinderer, T. E., K. W. Tucker, and A. M. Collins. 1982. Nest cavity selection by swarms of European and Africanized honeybees. *Journal of Apicultural Research* 21:98-103.

Roberts, W. C. 1944. Multiple mating of queens proved by progeny and flight tests. *Gleanings in Bee Culture* 72:255-259, 303.

Robinson, G. E. 1985. Effects of a juvenile hormone analogue on honey bee foraging behaviour and alarm pheromone production. *Journal of Insect Physiology* 31:277-282.

Robinson, G. E., B. A. Underwood, and C. E. Henderson. 1984. A highly-specialized water-collecting honey bee. *Apidologie* 15:355-358.

Root, H. H. 1980. Wax. In: *ABC and XYZ of Beekeeping*. E. R. Root, H. H. Root, J. A. Root, and L. R. Goltz, eds., pp. 651-666. A. I. Root, Medina, Ohio.

Rösch, G. A. 1927. Über die Bautätigkeit im Bienenvolk und das Alter der Baubienen. Weiterer Beitrag zur Frage nach Arbeitsteilung im Bienenstaat. *Zeitschrift für Vergleichende Physiologie* 6:265-298.

Rose, R. I. and J. D. Briggs. 1969. Resistance to American foulbrood in honeybees. IX. Effects of larval food on growth and viability of *Bacillus larvae*. *Journal of Invertebrate Pathology* 13:74-80.

Rosov, S. A. 1944. Food consumption by bees. *Bee World* 25:94-95.

Rothenbuhler, W. C. 1958. Genetics and bee breeding of the honeybee. *Annual Review of Entomology* 3:161-180.

Rothenbuhler, W. C. 1964. Behavioral genetics of nest cleaning in honeybees. IV. Responses of F_2 and back cross generation to disease-killed brood. *American Zoologist* 4:111-123.

Rothenbuhler, W. C. 1982. Semidomesticated insects: honeybee breeding. In: *Genetics in Relation to Insect Management*. M. A. Hoy and J. J. McKelvey, eds., pp. 84-92. Rockefeller Foundation, New York.

Rothenbuhler, W. C., J.M.S. Kulinčević, and V. C. Thompson. 1980. Successful selection of honeybees for fast and slow hoarding of sugar syrup in the laboratory. *Journal of Apicultural Research* 18:272-278.

Roubik, D. W. 1978. Competitive interactions between neotropical pollinators and Africanized honey bees. *Science* 201:1030-1032.

Roubik, D. W. 1980. Foraging behavior of competing Africanized honeybees and stingless bees. *Ecology* 61:836-845.

Roubik, D. W., S. F. Sakagami, and I. Kudo. 1985. A note on distribution and nesting of the Himalayan honey bee *Apis laboriosa* Smith (Hymenoptera: Apidae). *Journal of the Kansas Entomological Society.* In press.

Ruttner, F. 1956. The mating of the honeybee. *Bee World* 3:2-15, 23-24.

Ruttner, F. 1962. Drohnensammelplätze. *Bienenvater* 83:45-47.

Ruttner, F. 1968a. Systématique du genre *Apis.* In: *Traité de biologie de l'abeille,* Vol. 1. R. Chauvin, ed., pp. 1-26. Masson, Paris.

Ruttner, F. 1968b. Les races des abeilles. In: *Traité de biologie de l'abeille,* Vol. 1. R. Chauvin, ed., pp. 27-44. Masson, Paris.

Ruttner, F. 1968c. Insémination artificielle. In: *Traité de biologie de l'abeille,* Vol. 4. R. Chauvin, ed., pp. 181-197. Masson, Paris.

Ruttner, F. 1968d. Sexualité et reproduction. I. L'organe génital mâle et l'accouplement. In: *Traité de biologie de l'abeille,* Vol. 1. R. Chauvin, ed., pp. 145-185. Masson, Paris.

Ruttner, F. 1975a. Races of bees. In: *The Hive and the Honeybee.* Dadant and Sons, eds., pp. 19-38. Dadant and Sons, Hamilton, Ill.

Ruttner, F. 1975b. African races of honeybees. *Proceedings of the XXVth International Apicultural Congress, Grenoble,* pp. 325-344.

Ruttner, F. and K.-E. Kaissling. 1968. Über die interspezifische Wirkung des Sexuallockstoffes von *Apis mellifica* und *Apis cerana. Zeitschrift für Vergleichende Physiologie* 59:362-370.

Ruttner, F. and V. Maul. 1983. Experimental analysis of reproductive interspecies isolation of Apis-Mellifera L. and Apis-Cerana Fabr. *Apidologie* 14:309-328.

Ruttner, F. and H. Ruttner. 1965. Untersuchungen über die Flugaktivität und das Paarungsverhalten der Drohnen. 2. Beobachtungen an Drohnensammelplätzen. *Zeitschrift für Bienenforschung* 8:1-9.

Ruttner, F. and H. Ruttner. 1966. Untersuchungen über die Flugaktivität und das Paarungsverhalten der Drohnen. 3. Flugweite und Flugrichtung der Drohnen. *Zeitschrift für Bienenforschung* 8:332-354.

Ruttner, F., L. Tassencourt, and J. Louveaux. 1978. Biometrical-statistical analysis of the geographical variability of *Apis mellifera* L. I. Materials and methods. *Apidologie* 9:363-381.

Ruttner, H. 1976. Untersuchungen über die Flugaktivität und das Paarungsverhalten der Drohnen. VI. Flug auf und über Hohenrücken. *Apidologie* 7:331-341.

Ruttner, H. and F. Ruttner. 1972. Untersuchungen über die Flugaktivität und das Paarungsverhalten der Drohnen. V. Drohnensammelplätze und Paarungsdistanz. *Apidologie* 3:203-232.

Rutz, W., L. Gerig, H. Wille, and M. Lüscher. 1974. A bioassay for juvenile hormone (JH) and effects of insect growth regulators (IGR) on adult worker honeybees. *Mitteilungen Schweizerische Entomologische Gesellschaft* 47:307-313.

Rutz, W., L. Gerig, H. Wille, and M. Lüscher. 1976. The function of juvenile hormone in adult worker honeybees, *Apis mellifera*. *Journal of Insect Physiology* 22:1485-1491.

Sakagami, S. F. 1953. Untersuchungen über die Arbeitsteilung in einem Zwergvolk der Honigbiene. Beiträge zur Biologie des Bienenvolkes, *Apis mellifica* L. I. *Japanese Journal of Zoology* 11:117-185.

Sakagami, S. F. 1954. Occurrence of an aggressive behaviour in queenless hives, with considerations on the social organization of honeybee. *Insectes Sociaux* 1:331-343.

Sakagami, S. F. 1958. The false queen: fourth adjustive response in dequeened honeybee colonies. *Behaviour* 13:280-296.

Sakagami, S. F. and H. Fukuda. 1968. Life tables for worker honeybees. *Researches on Population Ecology* 10:127-139.

Sakagami, S. F., T. Matsumura, and K. Ito. 1980. *Apis laboriosa* in Himalaya, the little known world largest honeybee (Hymenoptera, Apidae). *Insecta Matsumurana* (New Series) 19:47-77.

Schaffer, W. M. 1974a. Selection for optimal life histories: the effects of age structure. *Ecology* 55:291-303.

Schaffer, W. M. 1974b. Optimal reproductive effort in fluctuating environments. *American Naturalist* 108:783-790.

Schaffer, W. M., D. W. Zeh, S. L. Buchmann, S. Kleinhans, M. Valentine Schaffer, and J. Antrim. 1983. Competition for nectar between introduced honey bees and native North American bees and ants. *Ecology* 64:564-577.

Schaller, G. B. 1972. *The Serengeti Lion*. University of Chicago Press, Chicago.

Schmid, J. 1964. Zur Frage der Störung des Bienengedächtnisses durch Narkosemittel, zugleich ein Beitrag zur Störung der Sozialen Bindung durch Narkose. *Zeitschrift für Vergleichende Physiologie* 47:559-595.

Schneidwein, E. M., H. Kala, B. Linzer, and J. Metzner. 1975. Zur Kenntnis der Inhaltsstoffe von Propolis. *Pharmazie* 30:803.

Schmid-Hempel, P. 1984. The importance of handling time for the flight directionality in bees. *Behavioral Ecology and Sociobiology* 15:303-309.

Schoener, T. W. 1979. Generality of the size-distance relation in models of optimal feeding. *American Naturalist* 114:902-914.

Scholze, E., H. Piechler, and H. Heran. 1964. Zur Entfernungsschätzung der Bienen nach dem Kraftaufwand. *Naturwissenschaften* 51:69-70.

Seeley, T. 1977. Measurement of nest cavity volume by the honey bee (*Apis mellifera*). *Behavioral Ecology and Sociobiology* 2:201-227.

Seeley, T. D. 1978. Life history strategy of the honey bee, *Apis mellifera*. *Oecologia* 32:109-118.

Seeley, T. D. 1979. Queen substance dispersal by messenger workers in honeybee colonies. *Behavioral Ecology and Sociobiology* 5:391-415.

Seeley, T. D. 1982a. Adaptive significance of the age polyethism schedule in honeybee colonies. *Behavioral Ecology and Sociobiology* 11:287-293.

Seeley, T. D. 1982b. How honeybees find a home. *Scientific American* 247(Oct):158-168.

Seeley, T. D. 1983. Division of labor between scouts and recruits in honeybee foraging. *Behavioral Ecology and Sociobiology* 12:253-259.

Seeley, T. D. 1985. The information-center strategy of honeybee foraging. *Fortschritte der Zoologie* 31:75-90.

Seeley, T. D. and B. Heinrich. 1981. Regulation of temperature in the nests of social insects. In: *Insect Thermoregulation*. B. Heinrich, ed., pp. 159-234. Wiley, New York.

Seeley, T. D. and R. A. Morse. 1976. The nest of the honey bee (*Apis mellifera*). *Insectes Sociaux* 23:495-512.

Seeley, T. D. and R. A. Morse. 1977. Dispersal behavior of honey bee swarms. *Psyche* 83:199-209.

Seeley, T. D. and R. A. Morse. 1978. Nest site selection by the honey bee. *Insectes Sociaux* 25:323-337.

Seeley, T. D., R. A. Morse, and P. K. Visscher. 1979. The natural history of the flight of honey bee swarms. *Psyche* 86:103-113.

Seeley, T. D., R. Hadlock Seeley, and P. Akratanakul. 1982. Colony defense strategies of the honeybees in Thailand. *Ecological Monographs* 52:43-63.

Seeley, T. D. and P. K. Visscher. 1985. Survival of honeybees in cold climates: the critical timing of colony growth and reproduction. *Ecological Entomology* 10:81-88.

Sekiguchi, K. and S. F. Sakagami. 1966. Structure of foraging population and related problems in the honeybee, with considerations on the division of labor in bee colonies. *Hokkaido National Agricultural Experiment Station Report* 69:1-65.

Sharma, P. L. 1960. Experiments with *Apis mellifera* in India. *Bee World* 41:230-232.

Shearer, D. A. and R. Boch. 1965. 2-Heptanone in the mandibular gland secretion of the honey-bee. *Nature* 206:530.

Sherman, P. W. 1980. The limits of ground squirrel nepotism. In: *Sociobiology: Beyond Nature/Nurture?* G. W. Barlow and J. Silverberg, eds., pp. 504-544. Westview, Boulder, Colorado.

Sherman, P. W. 1981. Kinship, demography, and Beldings ground squirrel nepotism. *Behavioral Ecology and Sociobiology* 8:251-259.

Silberrad, R.E.M. 1976. *Bee-keeping in Zambia*. Apimondia, Bucharest.

Simpson, J. 1950. Humidity in the winter cluster of honey-bees. *Bee World* 31:41-44.

Simpson, J. 1957a. The incidence of swarming among colonies of honey-bees in England. *Journal of Agricultural Science* 49:387-393.

Simpson, J. 1957b. Observations on colonies of honey-bees subjected to treatments designed to induce swarming. *Proceedings of the Royal Entomological Society of London* (A)32:185-192.

Simpson, J. 1958. The factors which cause colonies of *Apis mellifera* to swarm. *Insectes Sociaux* 5:77-95.

Simpson, J. 1959. Variation in the incidence of swarming among colonies of *Apis mellifera* throughout the summer. *Insectes Sociaux* 6:85-99.

Simpson, J. 1964. The mechanism of honeybee queen piping. *Zeitschrift für Vergleichende Physiologie* 48:277-282.

Simpson, J. 1973. Influence of hive-space restriction on the tendency of honeybee colonies to rear queens. *Journal of Apicultural Research* 12:183-186.

Simpson, J. and S. M. Cherry. 1969. Queen confinement, queen piping and swarming in *Apis mellifera* colonies. *Animal Behaviour* 17:271-278.

Singh, S. 1950. Behavior studies of honeybees in gathering nectar and pollen. *Cornell University Agricultural Experiment Station Memoir* 288:1-34.

Smith, F. G. 1953. Beekeeping in the tropics. *Bee World* 34:233-245.

Smith, F. G. 1958. Beekeeping observations in Tanganyika 1949-1957. *Bee World* 39:29-36.

Smith, F. G. 1960. *Beekeeping in the Tropics*. Longmans, London.

Snodgrass, R. E. 1956. *Anatomy of the Honey Bee*. Cornell University Press, Ithaca, N.Y.

Southwick, E. E. 1982. Metabolic energy of intact honey bee colonies. *Comparative Biochemistry and Physiology* 71:277-281.

Southwick, E. E., G. M. Loper, and S. E. Sadwick. 1981. Nectar production, composition, energetics and pollinator attractiveness in spring flowers in western New York. *American Journal of Botany* 68:994-1002.

Southwick, E. E. and J. N. Mugaas. 1971. A hypothetical homeotherm: the honeybee hive. *Comparative Biochemistry and Physiology* 40A:935-944.

Southwick, E. E. and D. Pimentel. 1981. Energy efficiency of honey production by bees. *Bioscience* 31:730-732.

Spangler, H. G. and S. Taber, III. 1970. Defensive behavior of honey bees toward ants. *Psyche* 77:184-189.

Starr, C. K. 1979. Origin and evolution of insect sociality: a review of modern theory. In: *Social Insects*, Vol. 1. H. R. Hermann, ed., pp. 35-79, Academic Press, New York.

Stearns, S. C. 1976. Life-history tactics: a review of the ideas. *Quarterly Review of Biology* 51:3-47.

Stort, A. C. 1974. Genetic study of aggressiveness in two subspecies of *Apis mellifera* in Brazil. 1. Some tests to measure aggressiveness. *Journal of Apicultural Research* 13:33-38.

Stort, A. C. 1975. Genetic study of aggressiveness in two subspecies of *Apis mellifera* in Brazil. IV. Number of stings in the gloves of the observer. *Behavioral Genetics* 5:269-274.

Stubblefield, J. W. 1980. Theoretical elements of sex ratio evolution. Ph.D. thesis, Harvard University.

Sturtevant, A. H. 1938. Essays on evolution. II. On the effects of selection on social insects. *Quarterly Review of Biology* 13:74-76.

Sturtevant, A. P. and I. L. Revell. 1953. Reduction of *Bacillus larvae* spores in liquid food of honeybees by the action of the honey stopper, and its relation to the development of American foulbrood. *Journal of Economic Entomology* 46:855-860.

Stussi, T. 1972. L'hétérothermie de l'abeille. *Archives des Sciences Physiologiques* 26:131-159.

Swain, T. 1977. Secondary compounds as protective agents. *Annual Review of Plant Physiology* 28:479-501.

Swammerdam, J. 1737. *Biblia Naturae*. Leyden.

Szabo, T. I. 1983a. Effects of various entrances and hive direction on outdoor wintering of honey bee colonies. *American Bee Journal* 123:47-49.

Szabo, T. I. 1983b. Effect of various combs on the development and weight gain of honeybee colonies. *Journal of Apicultural Research* 22:45-48.

Taber, S. 1955. Sperm distribution in the spermathecae of multiply mated queen honeybees. *Journal of Economic Entomology* 48:522-525.

Taber, S. 1980. A population of feral honey bee colonies. *American Bee Journal* 119:842-847.

Taber, S. and J. Wendel. 1958. Concerning the number of times queen bees mate. *Journal of Economic Entomology* 51:786-789.

Tanada, Y. 1967. Effect of high temperature on the resistance of insects to infectious diseases. *Journal of Sericultural Science of Japan* 36:333-340.

Taylor, O. R. 1977. The past and possible future spread of Africanized honeybees in the Americas. *Bee World* 58:19-30.

Taylor, O. R. and M. D. Levin. 1978. Observations on Africanized honey bees reported to South and Central American government agencies. *Bulletin of the Entomological Society of America* 24:412-414.

Thompson, D. W. 1942. *On Growth and Form*, Vol. 1. Cambridge University Press, Cambridge.

Thompson, V. C. 1960. Nectar flow and pollen yield in southwestern Arkansas. *Report of the Arkansas Agricultural Experiment Station* 9:1-38.

Thompson, V. C. and W. C. Rothenbuhler. 1957. Resistance of American foulbrood in honey bees. II. Differential protection of larvae by adults of different genetic lines. *Journal of Economic Entomology* 50:731-737.

Townsend, G. F. and E. Crane. 1973. History of apiculture. In: *History of Entomology*. R. F. Smith, T. E. Mittler, and C. N. Smith, eds., pp. 387-406. Annual Reviews, Palo Alto, California.

Trivers, R. L. and H. Hare. 1976. Haplodiploidy and the evolution of social insects. *Science* 191:249-263.

Trivers, R. L. and D. E. Willard. 1973. Natural selection of parental ability to vary the sex ratio of offspring. *Science* 179:90-92.

Tucker, V. A. 1970. Energetic cost of locomotion in animals. *Comparative Biochemistry and Physiology* 34:841-846.

Velthuis, H.H.W. 1970. Ovarial development in *Apis mellifera* worker bees. *Entomologica Experimentalis et Applicata* 13:377-394.

Velthuis, H.H.W. 1972. Observations on the transmission of queen substances in the honey bee colony by the attendants of the queen. *Behaviour* 41:105-129.

Verheijen-Voogd, C. 1959. How worker bees perceive the presence of their queen. *Zeitschrift für Vergleichende Physiologie* 41:527-582.

Villaneuva, V. R., M. Barbier, M. Gonnet, and P. Lavie. 1970. Les flavonoides de la propolis. Isolement d'une nouvelle substance bacteriostatique: la pinocembrine. *Annales de l'Institute Pasteur* 118:84-87.

Villaneuva, V. R., D. Bogdanovsky, M. Barbier, M. Gonnet, and P. Lavie. 1964. Sur l'identification de la 3,5,7-trihydroxy flavone (galangine) à partir de la propolis. *Annales de l'Institute Pasteur* 106:292-302.

Visscher, P. K. 1982. Foraging strategy of honey bee colonies in a temperate deciduous forest. M.S. thesis, Cornell University.

Visscher, P. K. 1983. The honey bee way of death: necrophoric behaviour in *Apis mellifera* colonies. *Animal Behaviour* 31:1070-1076.

Visscher, P. K. and T. D. Seeley. 1982. Foraging strategy of honeybee colonies in a temperate deciduous forest. *Ecology* 63:1790-1801.

Vollbehr, J. 1975. Zur Orientierung junger Honigbienen bei ihren 1. Orientierungsflug. *Zoologische Jahrbücher. Abteilung für allgemeine Zoologie und Physiologie der Tiere* 79:33-69.

Waddington, K. D. 1980. Flight patterns of foraging bees in relation to artificial flower density and distribution of nectar. *Oecologia* 44:199-204.

Waddington, K. D. 1981. Patterns of size variation in bees and evolution of communication systems. *Evolution* 35:813-814.

Waddington, K. D. 1982. Honey bee foraging profitability and round dance correlates. *Journal of Comparative Physiology* 148:297-301.

Waddington, K. D. and B. Heinrich. 1981. Patterns of movement and floral choice by foraging bees. In: *Foraging Behavior: Ecological, Ethological, and Psychological Approaches*. A. C. Kamil and T. D. Sargent, eds., pp. 215-230. Garland STPM, New York.

Waddington, K. D. and L. R. Holden. 1979. Optimal foraging: on flower selection by bees. *American Naturalist* 114:179-196.

Walker, E.P., F. Warnick, S. E. Hamlet, K. I. Lange, M. A. Davis, H. E. Uible, and P. F. Wright. 1975. *Mammals of the World*. Johns Hopkins Press, Baltimore.

Watanabe, H. and Y. Tanada. 1972. Infection of nuclear-polyhedrosis virus in armyworm, *Pseudaletia unipuncta* Haworth (Lepidoptera: Noctuidae), reared at a high temperature. *Applied Entomology and Zoology* 7:43-51.

Watson, L. R. 1928. Controlled mating in honeybees. *Quarterly Review of Biology* 3:377-390.

Weaver, N. 1956. The foraging behavior of honeybees on hairy vetch. I. Foraging methods and learning to forage. *Insectes Sociaux* 3: 537-549.

Weaver, N. 1957. The foraging behavior of honeybees on hairy vetch. II. The foraging area and speed. *Insectes Sociaux* 4:43-57.

Weaver, N. 1979. Possible recruitment of foraging honeybees to high-reward areas of the same plant species. *Journal of Apicultural Research* 18: 179-183.

Weipple, T. 1928. Futterverbrauch und Arbeitsleistung eines Bienenvolkes im Laufe eines Jahres. *Archiv für Bienenkunde* 9:70-79.

Weismann, A. 1893. The all-sufficiency of natural selection. *Contemporary Review* 64:309-338.

Weiss, K. 1962. Untersuchungen über die Drohnenerzeugung im Bienenvolk. *Archiv für Bienenkunde* 39:1-7.

Weiss, K. 1965. Über den Zuckerverbrauch und die Beanspruchung der Bienen bei der Wachserzeugung. *Zeitschrift für Bienenforschung* 8:106-124.

Wells, H. and P. H. Wells. 1983. Honey bee foraging ecology: optimal diet, minimal uncertainty or individual constancy? *Journal of Animal Ecology* 52:829-836.

Wells, P. H. and J. Giacchino, Jr. 1968. Relationship between the volume and the sugar concentration of loads carried by honey bees. *Journal of Apicultural Research* 7:77-82.

West-Eberhard, M. J. 1975. The evolution of social behavior by kin selection. *Quarterly Review of Biology* 50:1-33.

West-Eberhard, M. J. 1978. Temporary queens in *Metapolybia* wasps: non-reproductive helpers without altruism? *Science* 200:441-443.

Wheeler, W. M. 1911. The ant-colony as an organism. *Journal of Morphology* 22:307-325.

White, J. W., Jr. 1975. Composition of honey. In: *Honey: A Comprehensive Survey*. E. Crane, ed., pp. 157-206. Heinemann, London.

White, J. W., Jr., M. H. Subers, and A. I. Schepartz. 1963. Identification of inhibine, the antibacterial factor in honey, as hydrogen peroxide, and its origin in a honey glucose oxidase system. *Biochimica et Biophysica Acta* 73:57-70.

Wilde, J. de and J. Beetsma. 1982. The physiology of caste development in social insects. *Advances in Insect Physiology* 16:167-246.

Wille, A. 1958. A comparative study of the dorsal vessels of bees. *Annals of the Entomological Society of America* 51:538-546.

Williams, G. C. 1966a. *Adaptation and Natural Selection*. Princeton University Press, Princeton, N.J.

Williams, G. C. 1966b. Natural selection, the costs of reproduction, and a refinement of Lack's principle. *American Naturalist* 100:687-690.

Wilson, E. O. 1971. *The Insect Societies*. Harvard University Press, Cambridge, Mass.

Wilson, E. O. 1976. Behavioral discretization and the number of castes in an ant species. *Behavioral Ecology and Sociobiology* 1:141-154.

Wilson, H. F. and V. G. Milum. 1927. Winter protection for the honeybee colony. *Research Bulletin of the Wisconsin Agricultural Experiment Station* 75:1-47.

Wilson, W. T. 1971. Resistance to American foulbrood in honeybees. XI. Fate of *Bacillus larvae* spores ingested by adults. *Journal of Invertebrate Pathology* 17:247-255.

Winston, M. L. 1980. Swarming, afterswarming, and reproductive rate of unmanaged honeybee colonies (*Apis mellifera*). *Insectes Sociaux* 27:391-398.

Winston, M. L. 1981. Seasonal patterns of brood rearing and worker longevity in colonies of Africanized honey bees (Hymenoptera: Apidae) in South America. *Journal of the Kansas Entomological Society* 53:157-165.

Winston, M. L., J. A. Dropkin, and O. R. Taylor. 1981. Demography and life history characteristics of two honey bee races (*Apis mellifera*). *Oecologia* 48:407-413.

Winston, M. L. and S. J. Katz. 1982. Foraging differences between cross-fostered honeybee workers (*Apis mellifera*) of European and Africanized races. *Behavioral Ecology and Sociobiology* 10:125-129.

Winston, M. L., G. W. Otis, and O. R. Taylor, Jr. 1979. Absconding behaviour of the Africanized honeybee in South America. *Journal of Apicultural Research* 18:85-94.

Winston, M. L. and O. R. Taylor. 1980. Factors preceding queen rearing in the Africanized honeybee (*Apis mellifera*) in South America. *Insectes Sociaux* 27:289-304.

Winston, M. L., O. R. Taylor, and G. W. Otis. 1980. Swarming, colony growth patterns, and bee management. *American Bee Journal* 120: 826-830.

Winston, M. L., O. R. Taylor, and G. W. Otis. 1983. Some differences between temperate European and tropical African and South American honeybees. *Bee World* 64:12-21.

Wohlgemuth, R. 1957. Die Temperaturregulation des Bienenvolkes unter regeltheoretischen Gesichtpunkten. *Zeitschrift für Vergleichenden Physiologie* 40:119-161.

Woodrow, A. W. and E. C. Holst. 1942. The mechanism of colony resistance to American foulbrood. *Journal of Economic Entomology* 35:327-330.

Woyke, J. 1960. [Natural and artificial insemination of queen honey bees]. *Pszczelnicze Zeszyty Naukowe* 4:183-273. In Polish, with English summary. Summarized in *Bee World* 43:21-25.

Woyke, J. 1963. What happens to diploid drone larvae in a honeybee colony? *Journal of Apicultural Research* 2:73-75.

Woyke, J. 1964. Causes of repeated mating flights in queen honeybees. *Journal of Apicultural Research* 3:17-23.

Woyke, J. 1973a. Reproductive organs of haploid and diploid drone honeybees. *Journal of Apicultural Research* 12:35-51.

Woyke, J. 1973b. Experiences with *Apis mellifera adansonii* in Brazil and Poland. *Apiacta* 8:115-116.

Woyke, J. 1976. Population genetics studies on sex alleles on the honey bee using the example of the Kangaroo Island bee sanctuary. *Journal of Apicultural Research* 15:105-123.

Wright, S. 1933. Homozygosis and inbreeding. *Proceedings of the National Academy of Sciences, USA* 19:411-420.

Yokoyama, S. and M. Nei. 1979. Population dynamics of sex-determining alleles in honey bees and self-incompatability alleles in plants. *Genetics* 91:609-626.

Zeuner, F. E. and F. J. Manning. 1976. A monograph on fossil bees (Hymenoptera: Apoidea). *Bulletin of the British Museum of Natural History* 27:1-268.

Zimmerman, M. 1981. Patchiness in the dispersion of nectar resources: probable causes. *Oecologia* 49:154-157.

Zmarlicki, C. and R. A. Morse. 1963. Drone congregation areas. *Journal of Apicultural Research* 2:64-66.

Author Index

Adam, B., 14, 17
Adams, J., 21
Alexander, R. D., 5, 29
Alfonsus, E. O., 82
Alford, D. V., 46
Allen, M. D., 6, 18, 36, 38, 43, 45, 49, 53-54, 57, 64-65, 110, 112-113
Anderson, E. J., 112
Anderson, R. S., 111
Armbruster, L., 83
Avitabile, A., 24, 40-41, 43

Bailey, L., 111, 131-133
Bamrick, J. F., 134
Bartholomew, G. A., 108, 152
Bastian, J., 87, 109, 112-113, 121
Baxter, J. R., 18
Beetsma, J., 20
Betts, A. D., 103
Beutler, R., 19, 86, 92, 97, 103
Blum, M. S., 127, 134
Boch, R., 14, 55, 93, 96-99, 127
Bodenheimer, F. S., 43
Bonner, J. T., 155
Böttcher, F. K., 67-68
Bozina, K. D., 57
Breed, M. D., 55, 127
Briggs, J. D., 134
Brock, T. D., 110
Brokensha, D., 147
Brünnich, K., 82
Bulmer, M. A., 61-63, 66
Burgett, D. M., 134
Burleigh, R., 10
Butenandt, A., 134
Butler, C. G., 30, 60, 79, 89, 106, 126, 135

Cahill, K., 113-114
Cale, G. H., 16
Callow, R. K., 76, 134
Caron, D. M., 45
Chadwick, P. C., 110, 117
Chain, B. M., 111
Chandler, M. T., 145
Charnov, E. L., 26, 29, 50-53, 58, 63

Cherry, S. M., 64-65
Church, N. S., 153
Čižmárik, J., 132
Clutton-Brock, T. H., 139, 155
Cole, B. J., 5
Collins, A. M., 14, 127, 147
Combs, G. F., 76
Contel, E.P.B., 57
Craig, R., 25, 29, 51, 53
Crane, E., 14, 41, 142, 147
Crozier, R. H., 20, 25
Culliney, T. W., 9

Dadant, C. P., 23
Dade, H. A., 128
Daly, H. W., 141
Darchen, R., 77-78
Darwin, C. R., 5, 77, 89, 103
Dawkins, R., 5, 81
Demuth, G. S., 112-113
Dhaliwal, H. S., 150
Dold, H., 134
Dunham, W. E., 107, 118
Dyer, F. C., 32, 84
Dzierzon, J., 15

Eckert, J. E., 83, 86
Emerson, A. E., 5
Esch, H., 61, 74, 87, 108-109, 112-114, 121
Evans, H. E., 123

Farrar, C. L., 41, 68
Fell, R. D., 37, 45, 63, 76
Ferguson, A. W., 24, 31
Fisher, R. A., 50, 61
Fletcher, D.J.C., 138, 141-145
Fluri, P., 32, 143
Franks, N. R., 5
Free, J. B., 24, 31, 36, 45, 54, 86, 94, 102-103, 110, 112, 114-115, 126, 135
Friedmann, H., 148
Frisch, K. von, 18, 71, 83-87, 92-94, 96, 102, 126, 130, 145
Fukuda, H., 32, 82-83, 96-97, 143

Galton, D. M., 15
Gary, N. E., 18, 24, 30, 43, 67, 69-70, 86, 92, 131
Gates, B. N., 107, 112
Gauhe, A., 134
Getz, W. M., 6, 55-56, 60, 67
Ghisalberti, E. L., 132
Giacchino, J., 83, 97
Gilliam, M., 134
Glowska-Konopacka, S., 19
Gould, J. L., 14, 32, 72, 84, 86-88
Gowen, J. W., 16
Groot, A. P. de, 30
Guy, R. D., 142, 144, 147

Habermann, E., 129
Hamilton, W. D., 5, 25-26, 29, 51
Hannson, Å., 65
Harborne, J. B., 131
Hare, H., 25-26, 29, 52
Harvey, P. H., 139
Haslbachová, H., 23, 25, 30
Haydak, M. H., 82
Hazelhoff, E. H., 116
Hebling, N. J., 97
Heinrich, B., 40-41, 83, 92, 96, 100, 105-106, 108, 110, 113, 119-122, 145, 152
Heller, H. C., 40, 112-113, 115
Heran, H., 115
Herreid, C. F., 113
Hess, W. R., 107, 116
Heussner, A., 114
Himmer, A., 107, 110
Hirschfelder, H., 83
Hoefer, I., 14
Holden, L. R., 104
Hölldobler, B., 55
Holmes, W. G., 55
Holst, E. C., 131
Horn, H. S., 58
Horstmann, H.-J., 77
Huber, F., 38, 49, 57, 64, 77

Janscha, A., 15
Janzen, D. H., 123, 135
Jarman, P. J., 139
Jay, D. H., 22, 25
Jay, S. C., 20, 22, 24-25, 30, 110
Jaycox, E. R., 14, 72
Jeanne, R. L., 47
Jean-Prost, P., 57, 67

Jeffree, E. P., 43, 45
Johnson, D. L., 86
Jongbloed, J., 108, 113
Josephson, R. K., 108

Kaissling, K.-E., 11
Kalmus, H., 126
Kamil, A. C., 94
Kammer, A. E., 108, 120, 122
Katz, S. J., 14, 144
Kefuss, J. A., 45
Kerr, W. E., 20-21, 97
Kessel, B., 113
Kiechle, H., 117
Kigatiira, K. I., 147
Kloft, W., 128
Kluger, M. J., 111
Knaffl, H., 19, 86, 92
Koch, H. G., 40
Koeniger, G., 70, 138, 150
Koeniger, N., 70, 115, 138, 150
Köhler, F., 126
Koltermann, R., 14, 104
Korst, P.J.A.M., 6
Kosmin, N. P., 113
Krebs, J. R., 63, 94
Kronenberg, F., 40, 112-113, 115
Kropáčová, S., 22, 25, 30
Kulinčević, J. M., 16

Lacher, V., 115
Lack, D., 63, 71, 156
Laidlaw, H. H., 16, 21
Langstroth, L. L., 15
Lavie, P., 134
Lensky, Y., 18
Levchenko, I. A., 92
Levin, M. D., 19, 143
Lewontin, R. C., 5
Lindauer, M., 14, 32, 35-36, 61, 71-75, 78-80, 84, 95-97, 99, 101-102, 104, 116-118, 125, 154
Lindenfelser, L. A., 131
Little, H. F., 65
Lotmar, R., 133
Louveaux, J., 10, 45, 83
Lustick, S., 113-114
Lyr, H., 132

Maa, T., 10
Macevicz, S., 53

Mackensen, O., 16
Manning, F. J., 9
Marston, J. M., 70, 86
Martin, H., 78-79
Martin, P., 53, 60-61
Martins, E., 57
Maschwitz, U. W., 127-128
Matel, I., 132
Mau, R.F.L., 92
Maul, V., 10, 11
Maurizio, A., 103, 111
Mautz, D., 87
May, M. L., 152
McCleskey, C. S., 134
McLellan, A. R., 40
Melampy, R. M., 134
Menzel, R., 14, 104
Merrill, J. H., 82
Mestriner, M. A., 57
Metcalf, R. A., 21
Meyer, W., 60, 131
Michener, C. D., 11, 20, 23, 30-31, 46-47,
    55, 78, 123, 130, 140-141, 144
Milum, V. G., 36, 40, 110, 112
Mitchell, C., 63
Mitchell, W. C., 92
Mitchener, A. V., 40, 45
Morse, R. A., 17-18, 43, 53, 55, 67, 72,
    76, 124-125, 142, 144
Mugaas, J. N., 113
Murray, L., 45
Murrell, D. C., 153-154

Naile, F., 15
Nash, W. T., 153-154
Nedel, J. O., 31
Nei, M., 68
Neuhaus, W., 116
Neville, A. C., 108
Nightingale, J. M., 142, 145
Nitschmann, J., 133
Nolan, W. J., 43, 45, 82
Norton-Griffiths, M., 142
Nowogrodzki, R., 35, 97
Nunamaker, R. A., 141
Núñez, J. A., 97, 143-144
Nye, W. P., 16

O'Connor, R., 129
Oertel, E., 40, 97
Oettingen-Spielberg, T. zu, 99

Olifir, V. N., 92
Opfinger, E., 102
Orians, G. H., 94
Oster, G. F., 5, 25, 109, 146
Otis, G. W., 82, 96, 145, 148-149
Owens, C. D., 107, 112-113

Page, R. E., 21, 45, 49
Pager, H., 142, 147
Pain, J., 30
Parise, S. G., 14, 72
Park, O. W., 16, 83, 96, 99
Parker, R. L., 83, 94
Pearson, N. E., 94
Peck, M. L., 129
Peer, D. F., 68
Pellett, F. C., 15, 17
Pennycuick, C. J., 145
Percival, M. S., 97
Peters, R. H., 152
Phillips, E. F., 17, 112
Pimentel, D., 83
Pleasants, J. M., 104
Portugal Araújo, V. de, 144
Prica, M., 134
Pyke, G. H., 94

Queeny, E. M., 148

Ratner, S., 111
Rembold, H., 134
Revell, I. L., 132
Ribbands, C. R., 27, 36, 77, 79, 94, 99,
    103-104, 106, 126
Richards, K. W., 46
Riches, H.R.C., 129
Riemann, G., 133
Rinderer, T. E., 18, 72, 142-143
Roberts, W. C., 16, 20
Robinson, G. E., 32, 36
Root, H. H., 77
Rösch, G. A., 79, 126
Rose, R. I., 134
Rosov, S. A., 83
Roth, M., 114
Rothenbuhler, W. C., 15-17, 131-132, 134,
    136-137
Roubik, D. W., 12, 141
Ruttner, F., 10-11, 13-15, 20, 67-70, 140
Ruttner, H., 67-69
Rutz, W., 32

Sakagami, S. F., 6, 11, 24, 27, 32, 35-36, 82, 101, 125, 130, 143
Sargeant, T. D., 94
Schaffer, W. M., 58, 61, 93
Schaller, G. B., 142
Schmid, J., 102
Schmid-Hempel, P., 105-106
Schneidweind, E. M., 132
Schoener, T. W., 94
Scholze, E., 83, 96
Scovell, E., 5
Seeley, T. D., 17-19, 24, 30-35, 40-42, 45-46, 53, 58, 68, 72-74, 76-77, 79-80, 82-83, 86, 88-93, 95, 99-102, 110, 125, 136-137, 142, 144, 150, 156
Sekiguchi, K., 82, 101, 125
Sharma, P. L., 10, 150
Shearer, D. A., 127
Sherman, P. W., 55-56
Silberrad, R.E.M., 142, 144
Simpson, J., 18, 43, 45, 64-65, 107, 115
Singh, S., 36, 99, 106
Slabezki, Y., 18
Smith, F. G., 142, 144, 147
Smith, K. B., 56
Snodgrass, R. E., 23, 28, 70, 122, 128
Southwick, E. E., 40, 83, 96-97, 113-115, 134
Spangler, H. G., 126
Spencer-Booth, Y., 110, 112, 114
Starr, C. K., 25
Stearns, S. C., 61, 63
Stort, A. C., 14, 146
Stubblefield, J. W., 51
Sturtevant, A. H., 5
Sturtevant, A. P., 132
Stussi, T., 114
Swain, T., 131
Swammerdam, J., 15
Szabo, T. I., 73

Taber, S., 21, 68, 126
Tanada, Y., 111
Tassencourt, L., 10
Taylor, O. R., 111, 141-143, 145
Thompson, D. W., 78
Thompson, V. C., 40, 132
Townsend, G. F., 14
Trivers, R. L., 25-26, 29, 52-53
Tucker, V. A., 155

Veith, H. J., 115
Velthuis, H.H.W., 6, 24, 31
Verheijen-Voogd, C., 22
Villaneuva, V. R., 132
Vinson, S. B., 111
Visscher, P. K., 17, 19, 35-36, 40-42, 45-46, 68, 82-83, 86, 89-92, 95, 130
Vollbehr, J., 130
Voogd, S., 30
Vorwohl, G., 150

Waddington, K. D., 97-98, 104, 106
Walker, E. P., 148
Watanabe, H., 111
Watson, L. R., 16
Weaver, N., 36, 89, 103, 106
Weipple, T., 83
Weismann, A., 5
Weiss, K., 45, 49, 53, 77
Wells, H., 103
Wells, P. H., 83, 97, 103
Wendel, J., 21
Wenner, A. M., 86
West-Eberhard, M. J., 5, 25, 29
Whalley, P., 10
Wheeler, W. M., 5
White, J. W., 83, 134
Wiersma, C.A.G., 108, 113
Wilde, J. de, 20
Willard, D. E., 53
Wille, A., 122
Williams, G. C., 5, 61, 63
Williams, I. H., 45, 54, 86
Wilson, E. O., 5, 25, 31-32, 37, 109
Wilson, H. F., 112
Wilson, W. T., 132, 134, 141
Winston, M. L., 6, 14, 18, 38, 43, 49, 58, 63-64, 67, 139, 142, 144-146, 148-149
Witherell, P. C., 86
Wohlgemuth, R., 116
Woodrow, A. W., 131
Woyke, J., 20-21, 70, 141, 143
Wright, S., 68

Yokoyama, S., 68

Zeuner, F. E., 9
Zimmerman, M. L., 104
Zmarlicki, C., 67

Subject Index

absconding, 144-146
active space, pheromone, 24, 69
adaptation, study techniques, 4, 17-19, 138-139, 149-150
African honeybees, 139-149
afterswarms, 6-7, 37-38, 64-67
aggression: between colonies, 126-127; within colonies, 5-6, 27-28, 37-38, 49, 53
age polyethism: evolution, 32-36; pattern, 31-33
air currents, in nest, 73, 116-118
alarm communication, 127-128
allometry, 152-156
allozymes, 21, 57
alpine swift, *see Apus melba*
altruism: evolution of, 25-31; of individuals, 7, 22-31, 80, 125-126, 128-129
American foulbrood, *see Bacillus larvae*
ancestry of honeybees, 9-10
annual cycle of colonies, 39-48, 148-149
antennae, 24, 28, 78-79, 84-86, 115, 126
antennation, communication, 24, 97
antibiotics, 133-134
ants, as predators, 47, 126, 152, 156-158
apamin, 129
apiculture, history of, 4, 14-17
Apis: characteristic traits, 18-38, 139; distribution, 10-14; domestication, 14-17; fossils, 9-10; phylogeny, 10; species and races, 10-14, 140
Apis andreniformis, 11
Apis armbrusteri, 9-10
Apis cerana, 10-11, 13, 15, 139, 149-159
Apis dorsata, 11-13, 139, 149-159
Apis florea, 11-13, 136-137, 139, 149-159
Apis henshawi, 9
Apis laboriosa, 11-12
Apis mellifera: adansonii, see A. m. scutellata; carnica, 13-14, 17, 140; *caucasica*, 17; *ligustica*, 13-14, 17, 72-73, 140, 144; *mellifera*, 17, 140, 144; *scutellata*, 140-149
Apus melba, 148
artificial selection, 15-16
Ascosphaera apis, 111
auditory communication, 84-86

Bacillus larvae, 16, 130, 132, 134, 136-137
bactericidal agents, 111, 131, 134
bee-eater, *see Merops apiaster*
beeswax, *see* wax
birds, as predators, 148, 156-157
bivouacs, in swarming, 50, 74
Bombus: brood rearing, 39; colony cycle, 46-48; colony energy budget, 46-48; foraging, 100
bortniks, 15
breeding of honeybees, 15-17
brood: care of, 20-21, 29, 31-32, 36, 82, 107, 110-115, 131, 134, 136-137; production, 14, 41, 43-46, 82-83, 120, 146; social role, 24, 30, 113
broodnest, 32-35, 76, 107, 112-118
bumblebees, *see Bombus*
buzz-runners, 60-61, 74

cannibalism, 130, 146
caste: age, 31-36, 79; definition, 32; determination, 20, 30. *See also* division of labor
cells: drone, 53-54; queen, 36-37, 43, 64-65, 78; worker, 54, 76-79
central-place foraging, 80-106
chalkbrood, *see Ascosphaera apis*
chemical communication, *see* pheromones
chemical defense, 129, 131, 152, 157
chemoreception, *see* pheromones, taste
chilling, effect on brood, 107
cleaning of nest, 77, 130-131
climate: effect on annual cycle, 39-48, 148-149; effect on geographic distribution, 11, 141; effect on nest site, 14, 71-75, 112, 142
clustering: in swarming, 37, 61; in thermoregulation, 77, 109, 111-115, 143
cocoon, 65
coefficient of relatedness, 21-22, 25-28, 51-52, 54-57, 59-60, 61-63, 66
colony: density, 15, 17, 68; economics, 18, 39-43, 73, 76-77, 81-83, 100-103; foundation, 60-67, 71-79; growth, 43-46; life cycle, 36-38; mass, 40, 82-83, 150; odor, 55-56, 126-127; reproductive rate, 18, 36-

colony (*cont.*)
    38, 63-67, 148-149; survivorship, 38, 46,
    61-67
colony-level selection, 5-8, 80-81
colony size: evolutionary significance, 60-67,
    109-111, 130; in defense, 130, 135; in
    thermoregulation, 109-111, 120; measure-
    ments, 40, 82, 150
color learning, 103-104
combs, *see* nest
communication: of food sources, 18-19, 80,
    82-88, 101, 104; of intruders, 127-128; of
    nest overheating, 117-118; of nest sites,
    71-75; of presence of queen, 24, 30-31; of
    swarming, 60-61, 74; of water need, 117-
    118. *See also* antennation, dominance,
    pheromones, piping, recognition
comparative approach, 138-139
competition, reproductive, *see* aggression,
    kinship
compromise, among colony members, 49, 53
cooperation, 20, 25, 31, 49, 71-79, 80-81,
    107-121. *See also* communication, forag-
    ing, nest construction, thermoregulation
conflict, *see* aggression

dances, communicative: for food, 18-19, 80,
    82-88; for water, 118; on swarm, 71-75
dead, removal of, 35, 130-131
decision making: colony level, 49-67, 71-75,
    79, 93-103, 111-118, 135; individual
    level, 61-63, 97-99, 101-106
defecation, 132-133
defense, 123-137, 151-152, 156-159
demography, 32, 46, 61-67
desert, adaptation to, 121-122
despotism, 31
development: of colonies, 43-46; of individu-
    als, 36
digestive tract, 132-133
disease, 16. *See also Ascosphaera apis, Ba-
    cillus larvae, Nosema apis*
dispersal, colony, *see* migration, nest site se-
    lection, swarming
division of labor: among foragers, 35-36, 94-
    95, 99-101; among workers, 31-36, 94-95,
    99-101, 125-126, 130; reproductive, 20,
    22-31
domestication of honeybees, 14-17
dominance behavior, 29-31, 55

drone: comb, 53-54; congregation areas, 67-
    70; diploid, 21; investment in, 26, 29, 49-
    54, 59; mating behavior of, 67-70; parent-
    age, 22; production, 43-45, 49-54
dysentery, 132-133

eggs: laying by queen, 22, 27-28; laying by
    workers, 5-6, 22-25, 27-28, 30
Electrapis, 10
emigration, *see* absconding
endocrine glands, 20, 31-32, 143
endoparasitism, *see* disease
energetics: of colony, 40-43, 75-79, 81-83,
    100-103, 113-115, 120-121; of individual,
    94-99, 107-108, 152-156
excretion, *see* defecation
execution: of diploid drones, 21; of queens,
    57
exocrine glands, *see* hypopharyngeal, man-
    dibular, Nasanov, poison, wax

fanning, 116-118
feces, *see* defecation
feedback communication, 24-25, 53-54, 97-
    98
fever, 111
fitness, genetical, *see* inclusive fitness
flight muscles; as heat source, 107-109, 113-
    115, 121; as sound source, 60-61, 65, 74,
    84-86
flight speed, 96, 154-155
flower, constancy, 36, 103; learning, 14,
    103-104; scent, 86-87, 104, 126-127;
    searching, 87-88, 99-101, 103-106
food: collection, 80-106, 143-144; exchange,
    97-98; storage, 40-43, 79, 143-144. *See
    also* foraging
foraging: behavior of workers, 82-88, 93-
    106; colonial patterns of, 88-93. *See also*
    colony economics, communication, divi-
    sion of labor
foraging range, of colony, 18-19, 86, 89-92,
    150, 154-155
fossil honeybees, 9-10
fungicidal agents, 111, 131, 134

genetic load, 21
glands, *see* endocrine glands, exocrine
    glands
glucose oxidase, 134

group selection, *see* colony-level selection

growth of colonies, 43-46

guard duty, 35-36, 38, 65, 125-128, 156

hairless-black syndrome, 16

haplodiploidy: sex determination, 20; social significance, 20-31, 68. *See also* kinship

Helarctos malayanus, 156

hindgut, 132-133

history of apiculture, 4, 14-17

hives: development of, 14-16; social effects of, 18

homeostasis, *see* food, nest construction, thermoregulation

honey: consumption by colony, 40-43, 77, 82-83; storage of, 79, 134, 148

honey badger, *see Mellivora capensis*

honey-buzzard, European, *see Pernis apivorus*

honeyguide, *see Indicator indicator*

hormones, *see* endocrine glands

hornets, as predators, 156-157

10-hydroxy-2-decenoic acid, 134

hygiene, *see* nest cleaning

hypopharyngeal glands, 31, 134

hypothermia, 107, 110, 112, 121-122

identification signals, *see* recognition

inbreeding, mechanisms for avoiding, 51, 67-70

inclusive fitness, 5-8, 25-30, 54-67, 81

Indicator indicator, 148

individual-level selection, 5-8, 49, 80-81

inefficiency in colonies, 5-6, 80

information-center foraging, 81, 88-93

instrumental insemination, 16

integration, *see* communication, homeostasis

isolating mechanisms of species, 10-11

jostling behavior, 36, 130

juvenile hormone, 20, 31-32

kin recognition, 55-57, 66-67

kin selection, 25-31, 54-67, 81

kinship, evolutionary factor, 7, 25-31, 51-52, 54-57, 59-60, 61-63, 66-67

kin structure of colony, 20-22

labor, *see* caste, division of labor

language, *see* communication, dances

larvae: care of, 20-21, 29, 31-32, 36, 107, 110-115, 131, 134, 136-137; social role, 24, 30, 113

laying workers, *see* workers, laying

learning, 14, 32-36, 55-56

levels of selection, 5-8, 80-81

life history: of queen, 36-38, 57-60; of worker, 31-35

lizards, as predators, 156-157

local mate competition, 51, 67

local resource competition, 51-52

Macaca mulatta, 156

Malayan honey bear, *see Helarctos malayanus*

male, *see* drone

mammals, as predators, 147-148, 156-157

man: as predator, 14-15, 147-148, 156; as scientist, 4, 14-17

mandibles, social significance, 77-78, 126-127

mandibular glands, 30, 77, 127, 134

mating behavior, 20-22, 30, 67-70

Meliponini, 47

mellitin, 129

Mellivora capensis, 148

memory, *see* learning

Merops apiaster, 148

messenger bees, 24, 30

metabolic rate: of colony, 40-43, 113-115, 120-121; of individual, 83, 96, 107-108, 152

microclimate, *see* nest site, thermoregulation

migration of colonies, 138, 144-146, 150, 156, 158

mortality, *see* survivorship

multiple mating by queens, 20-22, 30, 69-70

Nasanov glands, 86

natural selection, units of, 5-8, 80-81

nectar: foraging for, 80-106; ripening, 134; storage of, 79, 134

nest: cleaning, 77, 130-131; construction, 76-79, 142; role in defense, 125, 136-137, 156-158; site selection, 48, 71-75, 136-137, 142, 151, 156-158; thermoregulation, 107-118; ventilation, 116-118; of *Apis cerana*, 11, 13, 150, 158; of *A. dorsata*, 11, 13, 150, 157; of *A. florea*, 11, 13, 150,

nest (*cont.*)
156-157; of *A. mellifera*, 11, 13, 71-79, 142
Nosema apis, 132-133
nuptial flights, 38, 67-70
nurse bees, 31-33, 78, 118, 134

odor, *see* colony odor, flower scent, pheromones
Oecophylla smaragdina, 152, 156-157
olfactory communication, *see* pheromones
orientation flights, 130
ovarian development in workers, 22-25, 30
oviposition, *see* eggs
9-oxo-2-decenoic acid, 7, 24, 30-31, 67, 69

parasites, 123-125
parental behavior, *see* brood care
parental manipulation, 7, 29-30, 63-64
Pernis apivorus, 156
phenotype matching, 55
pheromones: alarm, 127-128; brood, 115; colony odor, 55-56, 126-127; queen substance, 7, 24, 30-31, 67, 69; recognition, 55-57, 115, 130-131; recruitment, 86, 127-128; sex attractant, 24, 30, 67, 69
piping, queen, 65
planing behavior, 18
poison glands, 129
pollen: consumption by colony, 41, 82-83; foraging for, 80-106; storage of, 102-103
polyethism, *see* division of labor
population: effective size (genetic), 68; size of colonies, 43-46, 60-67, 82-83, 130, 150
predation, on colonies, 47, 63, 73, 141-142, 145, 146-149, 156-159
primates, as predators, 147-148, 156-157
propolis: as building material, 81, 112, 131-132; for colony defense, 131, 152, 157
protozoa, as parasites, 132-133
proventriculus, 132-133

queen: cells, 36-37, 43, 64-65; confinement, 64-65; court, 24, 30-31; distinctive features, 20, 22-25, 28; evolution of, 25-31, 47; fighting, 6, 57; investment in, 26, 29, 49-54, 59; lifespan, 20, 57-60; mating behavior of, 20-22, 30, 67-70; paternity of, 20-22, 54-57; piping, 65; production, 30,

36-38, 43-46, 49-54; signaling presence of, 30-31; substance pheromone, 7, 24, 30-31, 67, 69; supersedure, 38, 57-60; virgin, 37-38, 64-65, 67-70
queen excluder, 23
queenless colonies, 23-25

racial variation, 12-14, 111
ratel, *see Mellivora capensis*
recognition: of brood, 115; of intruders, 126; of kinship, 55-57; of nestmates, 126-127
recognition alleles, 55
recruitment communication, 18-19, 71-75, 80, 82-88, 101, 104
relatedness, *see* coefficient of relatedness, kinship, kin structure of colony
reproduction of colonies, *see* drone, swarming
reproductive behavior, *see*, mating behavior
reproductive effort, 60-63
resin, *see* propolis
rhesus monkeys, *see Macaca mulatta*
ritualization, 83-84
robbing, 126-127, 135
rock paintings, 14, 147
royal jelly, 20, 36

Schwirrlauf communication, *see* buzz-runners
scout bees, 71-75, 80, 95, 99-101
senescence of queens, 57-60
Serengeti ecosystem, 142, 145
sex determination, 20-21
sex ratio: investment, 26, 29, 49-54, 59; numerical, 49-50, 63
sexual behavior, *see* mating behavior
shaking communication, 36
signals, *see* communication
smell, *see* pheromones
sound communication, *see* auditory communication
sperm, 20-21, 69-70
sterility of workers, *see* worker, sterility of
sting, 127-129, 151, 157
stingless bees, *see* Meliponini
storage, *see* food storage
supernumerary queens, 6, 57
supersedure of queen, 38, 57-60
survivorship: of colonies, 38, 46, 61-67, 77, 156; of individuals, 32, 57-60, 82

swarming, 6-7, 18, 36-38, 43-48, 130, 144-146, 148-149
swarms: behavior of, 37, 53, 55, 60-61, 71-75, 145-146; composition of, 50, 60; season of, 43-48, 145; size of, 50, 60-67; thermoregulation in, 118-121
systematics of *Apis*, 10-14, 140

tactile: communication, 60-61, 65, 74, 97-98, 118; sense, 65, 78-79
taste, 97-98, 118
temperature control, *see* thermoregulation
temporal castes, *see* age polyethism
thermoregulation: in individuals, 121-122, 152-153; in nests, 41, 48, 107-118, 143; in swarms, 118-121
touch, *see* tactile communication, tactile sense
transport costs, 154-155
trap-lining, 104
tree shrew, *see* Tupaia glis
tropics, social evolution in, 46-48, 63, 138-159
Tupaia glis, 156

undertaker bees, 35-36, 130-131

venom, 129
ventilation of nests, 116-118
Vespa tropica, 156-157

waggle dance, *see* dances
wasps: as predators, 156-157; as prey, 47
water: collection, 36, 81, 118; in thermoregulation, 116-118, 122
wax: glands, 28, 77, 79; manipulation and use, 76-79, 156; production, 76-79
weaver ants, *see* Oecophylla smaragdina
wild colonies, 14-15, 17, 77, 141
winter survival, 39-48, 63, 73, 76-77, 95, 107, 141
work, *see* division of labor
worker: determination, 20, 30; distinctive features, 20, 22-25, 28; evolution of, 25-31, 47; jelly, 20; laying, 5-6, 23-24, 27-28, 30, 81; paternity of, 20-22, 57; sterility of, 20, 22-31, 81. *See also* caste, division of labor
worker size, ecological significance of, 96-97, 150-156

yeasts, 124

Library of Congress Cataloging in Publication Data

Seeley, Thomas D.
   Honeybee ecology.

   (Monographs in behavior and ecology)
   Bibliography: p.   Includes indexes.
   1. Honeybee—Ecology.   2. Insects—Ecology.
   I. Title.   II. Series.
QL568.A6S44 1985   595.79′9   85-42704
ISBN 0-691-08391-6 (alk. paper)
ISBN 0-691-08392-4 (pbk.)